Access to PDF files of chapters:
http://bioquest.org/microbescount/

Microbes Count!

Problem Posing, Problem Solving, and Peer Persuasion
in Microbiology

Edited by John R. Jungck, Marion Field Fass, and Ethel D. Stanley

Associate Editor: Emily Smith

Managing Editor: Virginia Vaughan

Editorial Assistant: Patti Robinson

The BioQUEST Curriculum Consortium
in conjunction with
The Microbial Literacy Collaborative

To access referenced materials
(programs, datasets, etc)
in this revised edition of
Microbes Count!
you must briefly register at:

http://www.bioquest.org/mc_access

where you will also be asked
to verify that you own
a copy of *Microbes Count!*

ENJOY!

Support for this project was provided, in part, by:

Howard Hughes Medical Institute
BELOIT: BioQUEST Enhances Learning Outwards and Inwardly with Technology

National Science Foundation
#0127498 BEDROCK: Bioinformatics Education Dissemination: Reaching Out, Connecting and Knitting Together.
#9952525 Lifelines Online: Accessible Investigative Biology for Community Colleges
#9619019 Partnerships for Advanced Computational Infrastructure
#9354813 The BioQUEST Curriculum and Learning Tools Development Project
#8163340 Development of a Comprehensive, Contemporary Microbiology Laboratory

Special Permission to use selected materials from

The BioQUEST Library VI was granted by ACADEMIC PRESS

A Project of

The BioQUEST Curriculum Consortium

Printed by Keystone Digital Press
West Chester, Pennsylvania

Library of Congress Cataloging-in-Publication Data
on request from the BioQUEST Curriculum Consortium

ISBN 0-9723211-0-1

Table of Contents

Modules by Topic

General	Medical	Environmental
Cell/Metabolism p. 27 p. 57 p. 85 p. 101 p. 251	*Immunology/Infection* p. 125 p. 137 p. 323 p. 327	*Industrial/Food* p. 53 p. 63 p. 85 p. 101 p. 235 p. 251
Molecular/Genetics p. 63 p. 121 p. 153 p. 159 p. 167 p. 207 p. 235	*Epidemiology* p. 203 p. 313 p. 331	*Agricultural/Horticultural* p. 153 p. 279 p. 299
Growth p. 33 p. 41 p. 73 p. 109 p. 251	*Medical Forensics* p. 129 p. 203 p. 263 p. 343 p. 347 p. 353	*Ecology* p. 93 p. 221 p. 229 p. 345 p. 307
Phylogeny/Evolution p. 129 p. 137 p. 181 p. 191 p. 203 p. 207		*Control* p. 267 p. 279 p. 285 p. 299

Modules by Tools and Resources

Bioinformatics	Visualization and Tools	Simulations	Wet Labs, Cases, and Tabletop
Phylogeny-Multiple Species Searching for Amylase p. 63 Proteins p. 181 Tree of Life p. 191 1 Cell, 3 Genomes p. 207 Visualizing Proteins p. 235 *Phylogeny- One Species* Molecular Forensics p. 129 HIV Evolution p. 137 West Nile Virus p. 203	*NIH Image Analysis* Modeling Mold p. 33 Shaped to Survive p. 73 Variable Variegation p. 229 *Visualization* Visualizing Proteins p. 235 Why Count WBCs? p. 327 Maps and Microbes p. 347 *Excel* Lytic Phage Growth p. 41 Plague Outbreak p. 285 *Diagnostics* *TB* Antibiotic Resistance p. 125 *AuntieBody* Immune response p. 323	*Epidemiology* Measles p. 313 Disease Spread p. 331 Vaccine p. 357 *Biota* Microbial Growth p. 109 Predator Prey p. 245 *Lateblight* Potato Blight p. 299 *Ecobeaker* Mold Fights Back p. 267 *MicroGCK* Know a Microbe? p. 27 Complements p. 159 Conjugation p. 167 *Extend Models* Wine p. 85 Biosphere p. 93 Stratified Waters p. 221 Pfiesteria p. 307 *Hereditary Molecule* p. 121	*Wet Labs* Modeling Mold p. 33 Shaped to Survive p. 73 Yogurt p. 101 Variable Variegation p. 229 Kimchee p. 251 *Cases* Farmer and the Gene p. 153 Bioterrorism p. 263 Citrus Canker p. 279 Souvenirs p. 343 *Tabletop Exercises* Sourdough p. 53 Scale of the Microbial World p. 57

The BEDROCK Project Approach to Bioinformatics Education

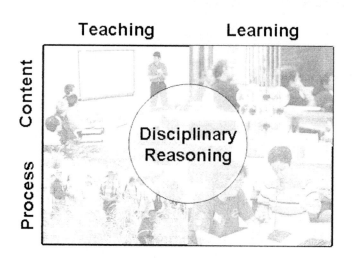

Figure 1. The BEDROCK focus on disciplinary reasoning blurs the distinctions between teaching or learning as well as content or process for the community of users engaging in a *Problem Space.*

Microbes Count! includes eight modules for investigating microbiology with bioinformatics resources developed with the BEDROCK Project (see facing page). BEDROCK *Problem Spaces* instantiate the pedagogical philosophy of BioQUEST's 3P's: Problem Posing, Problem Solving, and Persuading Peers. While a *Problem Space* defines an area of investigation and provides an initial data set and tools, students must still formulate the hypothesis that they will test, make decisions about a strategy of analyses that they will perform, and display their powers of communication rhetorically, aesthetically, and communally.

Problem Spaces more broadly consider the nature of problems and their solutions. Learners must move beyond the paradigm of simple problems with single unambiguous solutions to embrace a research model of science with creative opportunities and peer review defining the standard of significance and quality. Modules such as *Searching for Amylase* encourage users to seek creative solutions such as identifying a new microbial enzyme for industrial applications.

BEDROCK introduces bioinformatics by providing an area of investigation such as the variability of HIV strains within individual patients including an initial data set and tools within a *Problem Space.* Faculty members and students engage in focused inquiry using common data resources, access to analysis and visualization tools, reports of hypotheses, works in progress and findings. Knowledge claims are presented as the results of data-based arguments. As in any research community, users have opportunities to do new research and collaborate.

Each of the existing BEDROCK *Problem Spaces* provides an introduction, background on the area of study, collections of public data, information about and links to tools, bibliographic collections and curricular materials. These curricular materials act as resources for orienting new researchers (both faculty members and students) to the scientific community workspace.

Problem Spaces are designed to be open-ended and dynamic. Like the challenge of tracing the origin of the West Nile Virus NY99 strain, students have opportunities to grapple with analytical strategies for addressing emerging problems. *Problem Spaces* will provide opportunities to emphasize disciplinary reasoning and draw students into science instead of pushing them away.

The BEDROCK Project is a National Dissemination Project funded by the NSF Division of Undergraduate Education. BEDROCK is an acronym for Bioinformatics Education Dissemination: Reaching Out, Connecting, and Knitting-together. (For more information, see *bioquest.org/bedrock.*)

Descriptions of Bioinformatics Modules

Searching for Amylase

Keith D. Stanley and Ethel D. Stanley

An exploration of existing molecular databases and online bioinformatics tools to track down candidates for the next generation of industrial microbes.

p. 63

Molecular Forensics

Sam Donovan

This activity is based on an actual case which broke new ground with respect to the use of molecular evolutionary analysis of evidence in a courtroom.

p. 129

**Exploring HIV Evolution:
An Opportunity for Research**

Sam Donovan and Anton E. Weisstein

Sequence evolution over the course of infection with HIV is examined phylogentically within and between fifteen different AIDS patients.

p. 137

**Proteins: Historians
of Life on Earth**

Garry Duncan, Erick Martz,
and Sam Donovan

Conserved regions, such as active sites, within multiple sequence alignments are visualized on three dimensional visualizations of proteins.

p. 181

**Tree of Life: Introduction
to Microbial Phylogeny**

Beverly Brown, Sam Fan, LeLeng To Isaacs
and Min-Ken Liao

This activity familiarizes the student with the use of Internet-accessible bioinformatics tools, methods and data through an exploration of phylogenetic relationships and molecular markers.

p. 191

Tracking the West Nile Virus

Erica Suchann and Mark Gallo

In order to determine the likely origin and epidemiological spread in the US, students compare viral DNA sequences from West Nile infected mammals and birds

p. 203

**One Cell, Three Genomes: Evidence
for Endosymbiosis**

John R. Jungck, Sam Donovan,
and John M. Greenler

Students develop and test hypotheses about the endosymbiotic theory of organelles and both horizontal gene transfer and protein trafficking between chloroplasts, mitochondria, nuclei in eukaryotic cells.

p. 207

Visualizing Microbial Proteins

Ethel D. Stanley and Keith D. Stanley

Microbial enzymes serve as the focus of this student exploration of bioinformatics and visualization tools.

p. 235

Acknowledgements

The editors of *Microbes Count!* would like to acknowledge all of the contributors and software authors without whom this collection would not have been possible. Their names can be found in the list of contributors and in the activities they support.

Microbes Count! came to life during a committee meeting of the National Research Council on Information Technology (a subcommittee of the Committee on Undergraduate Science Education) in 1997. After several years of interaction with members of the microbiology education community at American Society for Microbiology national meetings and at the Gordon Research Conference on Undergraduate Microbiology Education, extensive writing began at the BioQUEST Curriculum Consortium workshop on microbiology education in Summer 2001. The participants at that workshop, many of whom became the authors of activities in this collection, worked hard to develop investigative, learner-centered activities that they could use in their classrooms. Their enthusiasm, experience, and vision propelled this project to look more deeply at many classroom topics. Obvious synergy between participants led to the inclusion of several new activities making use of bioinformatics.

We also want to thank the BioQUEST software developers who have shared their models, simulations, analytical tools, and approaches to form the heart of the project and Academic Press' extended permission for the use of software from *The BioQUEST Library* for *Microbes Count!*. Complete versions of the simulation modules are available in *The BioQUEST Library VI*, published by Academic Press.

As we neared the end of this project we received helpful guidance from our academic reviewers, Professor William Coleman of the University of Hartford, Professor Amy Cheng Vollmer of Swarthmore College, Professor Erica Suchman of Colorado State University and Professor Maura Flannery of St. John's University.

We deeply appreciate the constant support of Amy Chang of the American Society for Microbiology during a long developmental phase. The groundwork that preceded this project was largely due to the vision of the Microbial Literacy Collaborative of the American Society for Microbiology.

The activities of the BioQUEST Curriculum Consortium are supported by the Howard Hughes Medical Institute, the National Science Foundation, Education Outreach and Training–Partnership for Advanced Computing Infrastructure, and generous in-kind support from Beloit College. Our initial funding in microbiology education was from the Foundation for Microbiology from 1985 through 1987. The Annenberg Project of the Corporation for Public Broadcasting was the first major funder of the BioQUEST Curriculum Consortium.

Here at Beloit College we received skillful assistance from students who enhanced the project: Joshua Tusin, Joanna Kramer, Angela Hahn, Kari Roettger, and Emily Smith. Special recognition is extended to BioQUEST staffers who cheerfully took on emerging tasks. Our thanks to Tia Johnson, who refined methods and extended available data for key lab activities and to Sue Risseeuw, who organized the permissions for images in this collection. Tony Weisstein's superb mathematical insights and extensions to a number of activities in the flurry of pre-press activity deserve our special thanks.

Finally, we must emphasize our heartfelt appreciation to the Managing Editor of the BioQUEST Curriculum Consortium, Virginia Vaughan, whose extraordinary skills and shared vision brought this project to completion. *Our special thanks to Amanda Everse whose editorial and technical skills contributed to the publication of this revised 2006 version with corrections, new tables for accessing activities, and bioinformatics notes.*

The BioQUEST Curriculum Consortium

Authors

Julyet Aksiyote Benbasat
University of British Columbia

Spencer Benson
University of Maryland

Beverly Brown
Nazareth College of Rochester

Donald Buckley
Quinnipiac University

William Coleman
University of Hartford

Janet Decker
University of Arizona-Tucson

Jean Douthwright
Rochester Institute of Technology

Sam Donovan
Beloit College

Garry Duncan
Nebraska Wesleyan University

Samuel Fan
Bradley University

Marion Field Fass
Beloit College

Mark Gallo
Niagara University

John McC. Greenler
Beloit College

Robin McC. Greenler
Beloit College

Brooke Halgren
Beloit College

LeLeng To Isaacs
Goucher College

Tia Johnson
BioQUEST

John R. Jungck
Beloit College

K. C. Keating
Beloit College

Min-Ken Liao
Furman University

Eric Martz
University of Massachusetts- Amherst

Eli Meir
University of Washington

Elisabeth C. Odum
Santa Fe Community College

Howard T. Odum
University of Florida

Frank Percival
Westmont College

Amy Sapp
Harvard School of Public Health

Janet Yagoda
Shagam RhizoTech

Emily Smith
BioQUEST

Ethel D. Stanley
Beloit College

Keith D. Stanley
Tate & Lyle North America

Erica Suchman
Colorado State University

Joshua Tusin
Beloit College

Virginia Vaughan
BioQUEST

Janet Vigna
Southwest State University, Minnesota

Margaret Waterman
Southeast Missouri State University

Linda Weinland
Edison Community College

Anton E. Weisstein
BioQUEST

Peter Woodruff
Champlain St-Lambert College

Robert Yuan
University of Maryland

Software Developers

Phil Arneson, Cornell University
Lateblight

Karen Barrett, University of Hartford
Cell Differentials

Donald Buckley, Quinnipiac University
Cell Differentials, Search for the Hereditary Molecule

John N. Calley, Eli Lilly and Company
Microbial Genetics Construction Kit

Adrian Casillas, MD UCLA School of Medicine
Auntie Body IMMEX

Center for Polymer Studies Boston University
Fractal Dimension

Deborah Clark, Quinnipiac University
Cell Differentials

William Coleman, University of Hartford
Search for the Hereditary Molecule: Avery

Jim Danbury, University of Chicago
Biota

Will Goodwin, University of Oregon
Epidemiology

Lynn Gugiotti, University of Hartford
Cell Differentials

Benjamin Jones, Applied Biosystems
Biota

Renee Judd, Northwestern University
BGuILE: TB Lab

John R. Jungck, Beloit College
Microbial Genetics Construction Kit

John Kruper, University of Chicago
Biota

Richard Leider, Northwestern University
BGuILE: TB Lab

Jim Lichtenstein, University of Chicago
Biota

Eli Meir, University of Washington
Ecobeaker2

JoAnne Morrica, University of Hartford
Cell Differentials

Eric Nelson, University of Chicago
Biota

Elisabeth Odum, Santa Fe Community College
HypoxiaZone, Late Blight Life Cycle, SimBio2, Toxic Pfiesteria, Wine Fermentation

Howard T. Odum, University of Florida
HypoxiaZone, Late Blight Life Cycle, SimBio2, Toxic Pfiesteria, Wine Fermentation

Wayne Rasband, National Institutes of Health
NIH Image

Brian Reiser, Northwestern University
BGuILE: TB Lab

William Sandoval, Northwestern University
BGuILE: TB Lab

Jeff Schank, University of California-Davis
Biota

Mel Stave, Ulysses S. Grant High School
Auntie Body IMMEX

William Sterner, University of Chicago
Biota

Ronald Stevens, UCLA School of Medicine
Auntie Body IMMEX

Barr E. Ticknor, Cornell University
Lateblight

Daniel Udovic, University of Oregon
Epidemiology

Joyce Weil, University of Chicago
Biota

William Wimsatt, University of Chicago
Biota

Preface

Humans and their ancestors have never existed without the presence of microorganisms within them and around them. Yet, the microbial world is misunderstood by many. Just as the video series, *Unseen Life on Earth* presents visual introductions to key aspects of microorganisms, *Microbes Count!* contains activities for exploration of the microbial world.

These activities do much more than give "coverage" of all aspects of the study of microbes. They present new and varied learning environments, which afford teachers and students appreciation of microorganisms in a variety of ways.

First encounters of those studying microorganisms are generally negative: they are dangerous infectious agents. While this is certainly an important aspect of microbiology both medically and historically, microorganisms that are pathogens or potential pathogens are a very small number of those that are presently known to us. Indeed, it often comes as a surprise to the novice when told that human beings are guests in a predominantly microbial world and that microorganisms are required for our survival on this planet. Suddenly in recent headlines, microbes have always been an integral part of human evolution and social history. Two out of the three of the domains of different species living on earth are microbial. They have important roles in ecology, evolution, and diversity and as models for investigating all life processes. As such, they provide excellent materials for investigation and for investigative learning –"thinking materials."

The nature of microorganisms requires enumeration in order to study them. Rapid growth rates, a singular advantage for study, require methods for counting microbes and ways to distinguish living and dead cells in samples. Samples are often taken over short periods of time. Plate counts and other methods for counting bacteria provide excellent ways to introduce and instill the need for quantification. Students are confronted at once with the need to enumerate, duplicate and verify the validity of sample data. This encourages quantitative skills, which are a foundation for experimentation. Using these methods, novice learners can examine significant problems in biology, since there is much remaining to learn about the microbial world. Recent nucleic acid technologies have revealed that many microorganisms exist in widely diverse environments but cannot be grown in labs at present. These organisms and their interactions with each other and with larger organisms remain to be studied.

Students using quantitative experimentation experience the values of defining a problem, investigating that problem and designing appropriate experiments to examine that problem. The activities in *Microbes Count!* are designed for students to have these experiences. The key to each activity is not a survey or demonstration of various phenomena, but to provide a window which can stimulate a student's imagination and encourage them to experiment. Students are doing science and not merely observing science. They are active participants in the scientific process. This sharpens learning skills beyond use of one technique or technology. The ability to organize thoughts, to be imaginative, to design and to execute experiments will serve students long after the latest technical breakthrough is history. In today's world, technology is developing at a rapid pace. It is vital and pragmatic for students to become active learners prepared for a lifetime of new challenges in the workplace.

Today's students have instant access to vast amounts of data, and tools for rapid and quantitative handling of that data. The DNA base sequences of the human genome and that of a number of other species, most of which are microbial, have been determined and are readily accessible. The ability to access and use these materials will enhance student preparation for a life of learning. This technology means that students can ask questions not possible to ask a few years ago. They can study the relatedness of organisms and ask questions concerning

The BioQUEST Curriculum Consortium

the roles of specific genes in different species. Database mining of microbial genomes provides many opportunities for investigation. This clearly indicates that the traditional role of "teacher" is redefined. Teachers are facilitators of information access and its uses and are no longer information repositories. This means that teachers have exciting opportunities to make students active learners and to provoke imaginative and inquisitive natures.

Part of this student (and teacher) intellectual development is learning collaboratively. Individuals contributing in a group project can learn from others in the group and can be more productive. Teamwork is expected in the workplace and is effective in the classroom. We learn the value of social interaction for problem solving.

In my professional life as a microbiology educator at a small liberal arts university, I have experienced the value of communication with peers. Microbiology is a rapidly changing field; it is essential to keep up to date in the field. Therefore, most other microbiology educators and I take every opportunity for learning and for contact outside home institutions. In many cases (such as mine), we are the sole microbiologist at an institution. Before the formation of the microbiology education group, Group W of the American Society for Microbiology, I discovered the BioQUEST program. Here was a workshop like no other: development of curricula collaboratively, uses of newer technologies, meeting and sharing with other workshop participants, continuing collaborations beyond the workshop time frame and reshaping and continually improving ideas for new learning environments. These were exactly the kinds of efforts I advocated later on with other microbiology educators through the American Society for Microbiology. Ideas can be shared, reviewed by peers and tested at home institutions. The value of collaboration, communication, testing and re-testing and peer review in curriculum development is visible throughout the activities in *Microbes Count!*.

Many of these activities use computer technology. This technology provides opportunities for new learning environments and, at the same time, supports traditional ones. In addition to data mining of extensive databases, and the capability for quantitative manipulation of data, computers provide rich visual images for data gathering. Computer technologies help students improve organization skills and are invaluable as presentation tools. Computer simulations do not replace "hands-on" laboratory experiences. Rather, computer programs and activities can provide a more thorough understanding of the laboratory work. Computer programs can be designed to encourage student extensions of emulation through their own experimentation in laboratories. Computer simulations clarify the necessity for controls, for instance. Computer technology can reach visual learners in better ways.

The activities in *Microbes Count!* will welcome students to the world of the microbiology professional. The experiences of collaboration, clarity and organization of thoughts, as well as peer review of group experiments, all place the practice of microbiology clearly as a social activity. When students appreciate that microorganisms recognize no political or geographical barriers, they will see the importance of microbiology in society. Microbes must be considered if we are to live together in harmony. Recent headlines remind us of the potential for misuse of microorganisms and the need for ethical dialogues. Above all, learning to reason after fact-finding and appropriate use of the scientific method will give value to individual learners and will serve the global community far greater than the creation of factual and technical experts.

The theme of BioQUEST is the "3 P's": Problem posing, problem solving and peer persuasion. This program immediately excited me after my first encounter with it. I remain so to this day, having survived the realization of the changes it would require of me. I had been "professor, teacher, and expert" for years and now I was just another learner. It has been a joy to embrace these changes and to exploit the new opportunities afforded me as an educator. As a result of these changes, I have written inquiry-based experiments, modified older types of experiments, and actively involved students in lecture and laboratory. Keeping in mind the goals of student inquiry, I have adapted my style to accommodate a student-centered approach. Yes, I show students sterile techniques, but I also ask them to explain why things are done in this way. My students give small group microbiology education at the national level (through the American Society for Microbiology) began to take shape. Even as I write this, I am preparing new, inquiry-based laboratories and lecture session for a microbiology course. Next year, the undergraduate introductory biology course will be entirely inquiry-based.

Inquiry-based activities are available in *Microbes Count!*. Students will be able to explore microbial genomes in depth, compare DNA sequences of microorganisms, simulate epidemics, quantitate images, simulate ecosystems, model antibiotic resistance and repeat classical and modern experiments in microbial genetics. These simulations ask students to vary situations, parameters and conditions, thereby encouraging problem posing. As one familiar with this approach, I anticipate even more integration of these activities into my classes. If these activities and the approaches used for incorporating them into learning environments are new to you, then I welcome you to exciting, new and productive learning adventures!

Finally, microbiologists, as all biologists, recognize and appreciate the forces of change. This collection of activities in microbiology is a snapshot in time, a collection that will grow, evolve, and adapt to future developments. The power of this investigative approach is that it will sustain and encourage learning even as future changes occur.

William H. Coleman
West Hartford, Connecticut

Forward

Microbes Count! Problem Posing, Problem Solving and Peer Persuasion in Microbiology is an important new educational initiative that has resulted from collaboration between the American Society of Microbiologists (ASM) and BioQUEST. Its approach is consistent with a broader philosophy of biology education directed towards making the student a microbiologist, a lifelong learner and a citizen at a time of unprecedented globalization. Our biology graduates emerge from our educational institutions into a changed landscape. This new environment is characterized by a multiplicity of career tracks, a constantly changing body of knowledge, a highly diverse workforce and the globalization of science and technology. And yet very few faculty members take the time to reflect on how much the world has changed since they were undergraduates. They are faced with the immediacy of organizing their next course and their thoughts focus on the tasks that they need to do in the next quarter or semester. Teaching now more than ever requires us to dwell on three important issues: the characteristics of our science students today, the careers pursued by them after graduation, and their attrition rates. Many of our students are brighter and better trained than ever before. For those who graduate, surveys indicate that for those who are not in school, many of them are not working in laboratories but rather are in a variety of occupations for which they were not adequately prepared. Attrition in the sciences continues to be a major problem. Aside from poor preparation in high school, two major reasons for students dropping out of science are the poor quality of teaching and the boredom of pursuing courses that do not seem to bear any relevance to anything that they might want to do in their future lives. These problems are more acute for under-represented minorities who typically have had little exposure to the culture of science in their early life.

For more than a decade, I and my friend and colleague, Spencer Benson, have framed an educational strategy focused on two concepts: "A Virtual Workplace" and "A Journey without Maps". In the former, we have integrated major concepts with modules, tasks and laboratory experiments that mimic the activities that scientists carry out in their laboratories. These course constructs also try to provide an insight into the different careers that are pursued by biologists today. Two dramatic examples in *Microbes Count!* that reflect this approach are the ones that examine the use of amylases in the production of high fructose sweeteners and the critical problem of antibiotic resistance as it relates to tuberculosis. Students can and do recognize the relevance of such modules to modern society as well as to their own professional pursuits. Such educational resources can be exploited more fully when linked to a research experience. This can be done by posing a dilemma, e.g. a search for alternative raw materials for producing sweeteners, novel strategies for overcoming antibiotic resistance. Besides driving students to master the scientific facts and applying them to a different set of circumstances, it also teaches certain invaluable lessons:

- Defining a biological problem is dependent on the desired objectives

- There may be several feasible solutions rather than a unique one

- The solution(s) may be partially effective

- The solution may be dictated by non-scientific criteria, e.g. efficiency, ease of scale-up, risk, cost.

Our second approach, "Journey without Maps," centers on introducing diversity themes into biology courses. One major way in which this can be done is by modifying the content so that it does not center solely on the American experience. As I have often noted when we lecture on industrial microbiology, the example that is frequently used for fermentation is the making of yogurt. One could just as easily talk about the making of soy sauce, a far more complex process that uses mixed fermentation and mixed cultures of bacteria and yeasts. In economic terms, far more people use soy sauce and revenues from this industry are far larger than for yogurt. The use of cross-cultural materials has some important effects:

- It exposes the students to the concept of appropriate technology, defined as the use of products or processes that are compatible with the development and culture of a country. It is relatively simple to use the most advanced technology; it is far more complicated to review all the available science in order to find the one that is applicable in a foreign environment.

- Working on a cross-cultural topic is a great equalizer. Middle-class white students do not have an inherent advantage in such tasks. Frequently minority students are more comfortable and more successful when dealing with such topics.

- When dealing with many of the desperate problems of the developing world, students become aware that science does real things for real people in real time.

When taken together with the "Virtual Workplace," "Journey without Maps" established an educational construct that defines the relevance of the courses to life beyond the university walls. It is this sense of relevance and connectedness that motivates the students to master their scientific material.

Even if you are involved in interdisciplinary and cross-cultural topics in your courses, this is still a step removed from experiencing diversity. Our campus is highly diverse (~30% minorities) and we have pursued a strategy where a significant part of a class grade is dependent on team work. Furthermore, teams are selected by the instructor and teams are mixed by race/ethnicity, gender, academic major (and GPA when appropriate). This has been done with the honors seminars that are limited in size and have some of the best students in the university, but it has also been done in large, regular courses where academic performance is more heterogeneous. The most important aspect to mixed teams is that the group projects have to be complex and difficult. If the team is to do well, it will have to maximize the contribution of all of its members. Think of diversity not as a luxury but a necessity. Extensive evaluations are carried out in these courses using written questionnaires and in some cases through focus groups that are conducted by the university's Center of Teaching Excellence. There are a number of recurring themes in the student responses. If it had been left to them, they would never have selected such a team to work with. Now that they had done it, they felt that this was one of the most important things that they had learned, i.e., to work with people from diverse backgrounds. And that it was the very diversity in backgrounds and experiences that was a source of strength in carrying out their assignments.

Those that attempt innovation in undergraduate biology education fall generally into two categories: those that tried and failed (and doubt it can be done) or those that have been successful but do not address the difficulties in getting it to work. No one should doubt that the creation of teaching materials such as *Microbes Count!* and their effective use in class are hard work. But if quality education is the central mission of a college/university, we should be able to obtain the resources to pursue our objectives. At the same time, we should make every effort to streamline and make more efficient the way we run and manage our courses.

The tragic events of 9/11 have been emphasized as a matter of national security. And they are. They have also accentuated the controversy about globalization. Globalization is today's reality and it has real consequences. For nations like ours, endowed with abundant resources, technology, a skilled work force, and an entrepreneurial environment, globalization ensures our prosperity. It also provides a lock on science and technology that ensures success in the future. The same can be said for many of the other industrial nations. The rich just get richer. The exact opposite is true for the poor countries, they become excluded from this new world system, and this generates bitterness and resentment. And this is the breeding ground for terrorism and political fanaticism. As we teach our biology students, we should make them sensitive to our cornucopia of technological riches, and that the sharing of a modicum of this wealth might ensure a more equitable and peaceful world.

Robert Yuan
College Park, Maryland

From a Student's Perspective

Emily Smith worked with the BioQUEST Curriculum Consortium during her junior and senior years at Beloit College. She helped with both the 2001 and 2002 curriculum development workshops, and has been very involved in the organization of the Microbes Count! *project. She graduated in 2002 with a self-designed double major in Genetics, Bioethics, and Society and Psychology.*

For several years, I have been personally interacting with the BioQUEST pedagogy. In my college classrooms, in my office, in conversation with faculty from across the county and the world, I have debated, discussed, and implemented the underlying principles of problem posing, problem solving, and peer persuasion. Perhaps the best testament I can offer to its utility from a student's perspective is to assert that the proficiencies gained from working with BioQUEST in the classroom have enabled me to move easily between my roles as a student and as a BioQUEST staff member. The skills I used to design projects, implement them, and defend or explain them to my college peers were the same skills I used when working on a project with BioQUEST staff. The underlying skills that are fostered by the challenge of working with BioQUEST modules are far more important to a student's future success and development than any individual lesson or class.

BioQUEST materials are challenging; they represent a shift away from traditional conceptions of instruction. As a student, I had very mixed feelings about them, because I knew when a new project was introduced that it was going to require a serious effort on my part. It usually meant extra research, integration of a variety of information sources, exploration of complicated questions, and learning how to implement new tools. I spent a good deal of my time in college struggling with these projects. As a result, I gained confidence in my ability to face a challenge, I feel competent working with a number of applications and tools, and I have skills that I hope will make me attractive to a variety of employers and graduate programs. Being challenged to work hard and probe deeply into questions may not always have appealed at the time, but the experience certainly has benefited me greatly.

Much of the BioQUEST pedagogy is organized around the concept of collaborative work. This approach ensures that students stretch themselves, both in social interaction and in academic development. The best group work promotes leadership skills, models working environments, and allows students to depend on each other for support, motivation, and innovation. When a group of people with different backgrounds, different skills, and different goals sit down to work through a problem together, they will often lead each other down paths they never would have explored alone. Many times, it is the simple act of discussing the challenge, brainstorming, putting ideas into words that guide the way towards a deeper understanding of complicated problems. Ideally, collaboration does not happen only between class members, but encourages interaction between students, professors, community members, and other professionals. It is a disservice to students when they are sent the message that they can only learn from a professor.

One of the strengths of a BioQUEST module is the student ownership of the materials. When students pose their own questions, gather their own data, and follow it all the way through the analytic process, they become invested in the answers they find and the new questions that arise. This is an important motivational tool for students, many of whom have difficulty finding a connection to their schoolwork, particularly in introductory science classes. In essence, this process moves the activity from a task done to earn a grade, to a personal project with investment in the outcome. Whatever the students' special areas of interest or ability, they are able to find a connection to themselves and make the modules useful and pertinent for their needs and goals. As non-biology major, I was a novelty in my upper level biology classes, but I believe that I was still a valuable part of my project groups because a wide range of skills and perspectives is a benefit when working on BioQUEST materials rather than a hindrance.

Microbes Count! activities are designed to be open-ended enough to allow students to pursue their own interests or proclivities within the context of the activity. Students are encouraged to gather their own data and investigate its meaning and relevance. This entire process promotes a sense of ownership of the activities and their outcomes. Students can gear their own education towards subjects that apply to them and their lives. Naturally, not all activities are undertaken in perfect conditions, but when students are allowed flexibility within a topical structure and given freedom to choose the focus of their projects, they may tap into the internal motivation that is essential for true scholarship.

Critical to understanding the message of a BioQUEST education is the often-disquieting reality that most of the problems one encounters in life outside the classroom setting do not have one single or final answer. *Microbes Count!* activities reflect this complexity, encourage diversity in understanding and answers, and rarely ask for the "correct" explanation. These modules are a move away from binary systems, from simple justifications, from one-sided arguments. This movement is reflected not only in the kinds of questions posed (or prompted), but also in the more general focus of the book on the ecology, evolution, and diversity of the microbial world. Too often, microbes are categorized strictly as pathogens or decomposers, or anthropomorphized as 'the enemy' or 'the pest'. Studying or discussing microbes on such human terms is certainly easier than fuller investigation of the microbial reality in the short run, but ultimately misses the opportunity for more enlightening understanding. Taking this broader approach to education is one important step towards a society of intelligently skeptical citizens and scientists. This is particularly important when one considers that the exposure to biology and scientific methodology for many students is quite limited. One or two courses may be the extent of their opportunity to acquire many of the critical thinking skills necessary for making informed decisions, not only in a research or scholarly environment, but in everyday life. This world community needs a population capable of interpreting information gathered from scientists, politicians, and media, not simply digesting it without discriminating thought.

Emily Smith
Beloit College

Introduction

Microbiology stands at the forefront of educational reform in Biology. Microbiologists have forged ahead with identification of curricula and the endorsement of active learning strategies. Each year the Undergraduate Education Conference that precedes the annual meeting of the American Society of Microbiology (ASM) introduces educators to new ideas and practices to enhance student learning.

With *Microbes Count! Problem Posing, Problem Solving, and Peer Persuasion in Microbiology*, the BioQUEST Curriculum Consortium has developed a series of activities to complement ASM's video series, *Unseen Life on Earth. Unseen Life on Earth* celebrated the many roles that microbes play in our environment, and we have used that approach to guide the selection of activities. This is not a set of activities about pathogens, but rather a set of activities that explore microbial growth, microbial metabolism and microbial interactions. As reflected in the title, *Microbes Count!*, we have tried to emphasize quantitative analyses in the activities we have included. We have used many computer simulations developed over the past 20 years by members of the BioQUEST Curriculum Consortium to address real problems in microbiology. We have moved as well into the use of bioinformatics to address questions of concern to microbiologists.

Several ideas have guided the development of this collection. We have first tried to follow the organization of the *Unseen Life on Earth* video series. This series moves students from a consideration of the microbial universe to an introduction to genetics, microbial interactions and ecology, and finally to concepts of microbes and disease. While we have used the video series to shape the organization of activities, we have not been restricted to the topics introduced and to the level of analysis implied. Some of the activities in *Microbes Count!* are clearly designed for the introductory student in an introductory class, while others take the student far past introductory ideas. We hope that you will use them as a springboard for student research.

In each chapter of *Microbes Count!*, we have tried to include activities at three levels–an introductory level that introduces basic concepts and skills, an intermediate level that provides some opportunity for open ended investigation and an advanced level that encourages students to integrate concepts and boldly use data in independent investigations. Advanced level activities typically extend over several laboratory periods, or may engage students out of class for several weeks.

Many of the investigations are suitable for non-majors, while others fit better in upper-level majors courses. We hope that this book will be used to complement wet labs and enhance on line courses, but we do not expect any one professor to use each activity or each chapter.

Although we had first envisioned a unified laboratory manual, we have realized as we put this together than we wanted to keep the voices and more importantly the approaches of the professors who contributed to this collection. These voices reflect the different ways that we approach the classroom. The BioQUEST Curriculum Consortium advocates learner-centered, problem-driven educational strategies, yet we recognize that there is not one best way to approach students. Some professors are more comfortable with highly structured activities that lead students to explore open ended questions, while others prefer to expose students directly to the issues and allow them to develop their own approaches to problem solving. The approach that you choose to adopt is an issue of personal preference.

All of the activities in this manual come from the classrooms of the developers, where they have been used with undergraduate students. Some of these activities have been used at large universities, while others have been used at small colleges. They have been tried at 2- year colleges and 4-year universities. Each is included

All of the activities in this manual come from the classrooms of the developers, where they have been used with undergraduate students. Some of these activities have been used at large universities, while others have been used at small colleges. They have been tried at 2- year colleges and 4-year universities. Each is included on the accompanying CD in a format that will allow faculty to adapt it to their needs.

Although the majority of activities in *Microbes Count!* are based on computer simulations, there are also activities that use bread mold, yogurt and kimchee as systems for analysis. We have included these because we believe in the value of hands-on laboratories and the benefit of using naturally occurring, non-pathogenic species in the lab (and in the kitchen for distance learners). In these activities we have also included data sets and photos so that students may use the data previously collected in their analyses.

Please review these activities to find the ones that challenge you to look differently at the microbial universe.

We are all learners as we explore graphic analysis with NIH image, or see where bioinformatics analyses take us in understanding evolutionary relationships. The activities in *Microbes Count!* are designed to provide rich problem spaces where faculty and students together explore the unsolved questions of microbiology.

Marion Field Fass
Beloit, Wisconsin

Chapter 1

Activities for Video I: The Microbial Universe

Peering through the microscope at our first slide of bacteria and seeing those small dots with hard-to-distinguish shapes, we soon realize that any reliable identification of microbes requires alternative methods. The activities in this chapter introduce strategies for identifying microbes, including the rationale for describing microbial growth quantitatively to help further characterize and understand the microorganisms in the world around us.

In this unit, you can:

- explore classic laboratory methods using a computer simulation to grow and characterize these minute life forms,
- quantify the effects of nutrient availability on microbial growth rates using visualization tools in the lab, and
- explore a multiple parameter model of population growth in bacteriophages that demonstrates interdependence with the population of bacterial hosts.

How Do You Know a Microbe When You Find One?

What are the distinguishing traits of microorganisms? Can we take advantage of the tendency of bacteria to form colonies on media in the laboratory to differentiate between mixed strains? The simulation *Microbial Genetics Construction Kit (μGCK)* introduces classic lab strategies such as serial dilution and replica plating, setting the stage for more exploration. Given an unknown sample of bacteria, can you devise strategies to identify the nutritional phenotypes of the microorganisms in your sample?

Modeling More Mold

How do we determine the pace of fungal growth on a limited substrate? Our medium here, a piece of white bread, provides easily observable data. Using digital cameras and image analysis tools, we can quantify the rate of growth and determine the mathematical functions that explain it. The application of mathematics to simple biology problems introduces students to some important tools for more in depth analyses.

Population Explosion: Modeling Phage Growth

How is the relationship between phages and their bacterial hosts like that of predators and prey? How does this relationship differ? What are the factors that contribute to phage production in the cell? What can the phage concentration in a high titer lysate tell us about both of these populations? We can investigate these questions and more with an Excel spreadsheet model for phage growth.

How Do You Know a Microbe When You Find One?

Marion Field Fass and Joshua Tusin

Video I: The Microbal Universe

Microbes are everywhere, but microbes underwhelming. They're small (most bacteria are only 1-5 microns in length) and hard to describe, even when using powerful light microscopes to see them.

There are several ways of describing bacteria, viruses, and other microbes. What sorts of characteristics could you use to describe microbes you can't see?

Figure 1. Morphology of Borrelia burgdorferi.

Introduction

One common way to characterize bacteria is by the kinds of diseases they cause. For instance, the *Borrelia burgdorferi* in Figure 1 causes Lyme Disease. The problem with this approach is that most microbes don't cause disease; in fact many have traits that are beneficial to the environment.

Another technique is to characterize microbes by the nutrients they use for energy and the by-products they produce, or by their preferred habitats. Think about the ways we describe different plants and animals. We classify all plants as photosynthetic, using light to produce food. Animals can be categorized by how they obtain food or by the kind of food they eat–carnivore, herbivore, or, for most humans, omnivore. We can also describe desert plants and rainforest plants in addition to desert animals or rainforest animals.

Bacteria also vary by habitat, by nutrient source, and by their susceptibility to certain antibiotics. These characteristics are coded for in the bacterial DNA. In mammals, DNA codes the hemoglobin that carries the oxygen in blood and the amylase that helps digest food. In bacteria, DNA similarly codes for the proteins that allow them to live in specific environments and to utilize specific nutrients.

If you have viewed the video series Unseen Life on Earth, you have seen bacteria that live in extremely high temperatures, like *Thermatoga*, and others that prefer to live in the rumen of a cow. For these bacteria identification by their habitat makes sense. For many bacteria, however, habitat differences are not so obvious and we have to look for other, less obvious ways to identify them.

Identifying bacteria

In this activity you will use the *Microbial Genetics Construction Kit* (*MicroGCK* or *μGCK*) to describe unknown bacteria by identifying their nutritional requirements. *MicroGCK* is a computer program that simulates a microbial genetics laboratory. You will have access to the same tools and procedures that a microbiologist would use to identify actual bacteria. You will be able to do serial dilution, create growth media containing specific nutrients, and do replica plating as if you were in a laboratory. You can inoculate plates, grow out colonies, and change the media on which these virtual bacteria grow.

You are given a "virtual" tube containing several kinds of unknown bacteria in broth. You don't have a microscope, but you have to describe the mixed bacterial broth as best as you can using the tools available to you.

In this activity you will need to determine:

- how to grow individual colonies of bacteria so that you can identify them, and
- how to vary the nutrients to help you describe these bacteria.

Getting Started

You start an experiment in *MicroGCK* by choosing a problem from the opening screen. (Detailed instructions for using *MicroGCK* are available in the Users Manual on the *Microbes Count!* web site http://bioquest.org/microbescount.)

For this activity, select "Serial Dilution/Phenotype Identification" and click on Start Problem. The first thing you will see is the Field Tube, the virtual tube of mixed bacteria. Initially the tube is gray, which means it contains millions of tiny bacteria, so many that the broth in the tube is opaque.

Bacteria are grown on petri plates, small dishes filled with a gelatin-like agar. Normally, the agar contains the nutrients that most bacteria need to grow. It is also much easier to count bacteria when they are grown on a petri plate, instead of in a test tube.

To inoculate a petri plate with the bacteria from your Field Tube, first select the tube by clicking on it. The name "Field Tube" will be highlighted when the tube is selected. In the Tube menu, choose the option Plate Tube. This operation will take 0.1 ml of bacterial broth from the Field Tube and spread it on a petri plate, incubate the plate, and then display the result for you. The nutrients on the plate will be the same as the ones in the tube.

Inoculating a plate with bacteria from your tube is simple, but how much bacterial broth is enough? When 0.1 ml of the bacterial broth from the Field Tube is spread out on the agar plate, the result is probably a plate that is entirely gray.

Figure 2. Screen shot from *MicroGCK*: Inoculating a plate with unknown bacterial broth resulting in a "lawn".

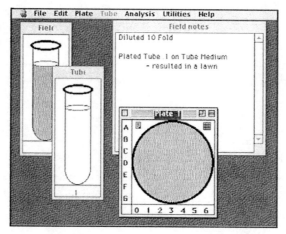

Microbiologists describe this as a "lawn," which means there is an even covering of bacteria which has no individual colonies. We can see a lawn on Plate 1 of Figure 2.

Remember that in order to describe the different kinds of bacteria in your Field Tube, you need to be able to grow individual colonies. What should you do next? Serial dilution is a process that takes your vial with millions of bacteria and systematically lowers the concentration of bacteria in that tube until you can plate out individual bacterial colonies.

Serial Dilution in *MicroGCK*

To perform a serial dilution in *MicroGCK*, make sure that the Field Tube is selected. (Remember, the tube is selected when its title is highlighted.) Now choose Dilute 10 Fold from the Tube menu. A new tube (Tube 1) is created with a broth that is nine tenths sterile water and one tenth your original bacterial broth.

This new tube has one tenth the population of the original and may be less gray than the first tube because there are fewer bacteria. If it is not less gray, do another dilution by selecting your new, diluted tube, and Dilute 10 Fold again. Continue doing dilutions until your tube looks white, or nearly white. (Be sure to have the correct tube selected when you choose to Dilute 10 Fold.)

Once you have a nearly clear tube, select that tube and once again choose Plate Tube from the Tube menu. This will inoculate a petri plate, incubate it, and display the results for you. Hopefully you will see individual bacterial colonies (gray circles) on your petri plate, as in Figure 3.

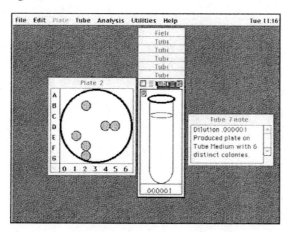

Figure 3. Screen shot from MicroGCK: Successful serial dilution results in Plate 2 with six distinct colonies.

If your plate is completely white, without any colonies, try plating a tube that is less diluted. If this results in a lawn, dilute the tube again and try plating this tube.

Counting the Bacteria

To determine the number of bacteria in the original tube, first count the number of colonies on your plate. Let's assume that there are six, like on Plate 2 in Figure 3. Now count how many times you diluted the original tube; we'll use six times, like in Figure 3. The important thing to remember is that you also did one last dilution when you plated the bacteria. You only used one tenth of your final dilution to make the plate. So, this means that your original tube of bacteria has $6 \times 10 \times 10^6$, or 60 million bacteria!

Now that you have finished the serial dilution, it is time to describe the bacteria you have isolated.

Phenotype Identification

In *MicroGCK*, each of the colonies on a plate are probably from different bacteria. Each colony represents a clone, or a group of bacteria that are all genetically the same. Thus, all of the bacteria in a single colony will have the same nutritional phenotype and antibiotic sensitivities–that is, they will all require the same nutrients in order to grow, and will be sensitive to the same antibiotics. Bacteria from different colonies will likely have different phenotypes.

To be able to analyze these bacteria, we must develop exact replicas of the plate so that we are able to track each colony and its nutritional needs. This process is known as replica plating.

To make an identical copy of your plate, select the plate you want to duplicate by clicking on the title of the plate. (If you click on a colony or a button you will select that colony or button and not the plate.) Under the Plate menu, select the Replicate option. This displays a window called Media Matrix where you can select the nutrients and/or antibiotics to be included in your new plate.

The Media Matrix window shows the title of the plate you are replicating and the nutrients included in the agar for that plate. There will also be a column for your new plate, with no nutrients selected. You will want to decide which nutrients to include in your replica plate so that you can learn something about your bacteria. It might be a good idea to only change one thing at a time, but you can choose your own method.

To include a nutrient or an antibiotic on the new plate, click the box for it. Once you have created a media that includes the nutrients you want, click the Plate button in the upper left. This will inoculate your new plate and incubate it, displaying the results. You can then compare the colonies that grow on the new media with those from your original plate.

If a colony still exists on the new plate, it means that the nutrients you removed from the plate were not necessary for that particular bacterial colony. However, if a colony is missing on the new plate, you have just learned something about which nutrients are required for it to grow.

As you experiment, you will want to devise a system to keep track of what you have learned about each colony. The Media Matrix maintains a record of the nutrients in each plate, but you should also keep track of what you know about each colony. You can use the Notebook available in *MicroGCK* (click on the small notebook icon in the upper left of each plate or tube window.)

Make sure to repeat your experiments so that you can confirm your results on new plates. Each problem in *MicroGCK* is unique and there is no answer book. You have to use the results from your experiments to support your analysis of the bacterial colonies.

Additional information on using the *Microbial Genetics Construction Kit* is available on the *Microbes Count!* web site at http://bioquest.org/microbescount.

- What advantage is it to make replica plates of your test plate, rather than just use a new batch of bacteria from the field tube for each new experiment?

- Each problem generated from *MicroGCK* is different. You will need to develop a strategy to justify your results and to convince others that your plan of experiments adequately demonstrates your conclusions. Write a report or develop a poster that describes the nutritional requirements and antibiotic sensitivities of the bacteria you have studied.

- What do the results of the replica plating tell you?

Software Used in this Activity

Microbial Genetics Construction Kit

John N. Calley (Eli Lilly and Company) and John R. Jungck (Beloit College)

Platform Compatibility: Macintosh only

Additional Resources

Available on the *Microbes Count!* web site http://bioquest.org/microbescount

Software

Microbial Genetics Construction Kit

Text

A PDF copy of this activity, formatted for printing

"Using the *Microbial Genetics Construction Kit*"

Related *Microbes Count!* Activities

Chapter 3: The Living World of Yogurt

Chapter 5: Complements Please! No Compliments Necessary

Chapter 5: Conjugation and Mapping

Unseen Life on Earth Telecourse

Coordinates with Video I: The Microbial Universe

Relevant Textbook Keywords

Antibiotics, Inoculum, Phenotype, Replica Plating, Serial Dilution

Related Web Sites (accessed on 2/21/03)

American Society for Microbiology
http://www.asmusa.org

Microbes Count! Website
http://bioquest.org/microbescount

Unseen Life on Earth: A Telecourse
http://www.microbeworld.org/htm/mam/is_telecourse.htm

References

Calley, J. N. and J. R. Jungck (2001). Microbial Genetics Construction Kit. In *The BioQUEST Library Volume VI.* Jungck, J. R. and V. G. Vaughan, Editors. Academic Press: San Diego, CA.

Figure and Table References

Figure 1.   Courtesy of Dr. Jeff Nelson, Rush University, Chicago, IL, USA
http://www.microbelibrary.org/images/jnelson/Images/borrelia.jpg

Figure 2.   Screen shot from *Microbial Genetics Construction Kit*

Figure 3.   Screen shot from *Microbial Genetics Construction Kit*

Modeling More Mold

John R. Jungck, Tia Johnson, Anton E. Weisstein, and Joshua Tusin

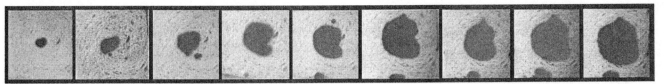

Figure 1: A partial data set for the bread mold *Mucor hiemalis* (+). Days 2–10 are shown; the area of mold growth has been been darkened for emphasis.

Exponential growth is so closely associated with microbiology that it is easy to imagine being waist deep in *E. coli* after only a few days of global warming. Yet it can be a challenge to understand what exponential growth actually means. In the "How Do You Know a Microbe When You See One" activity in Chapter 1, the growth of bacteria in a test tube is indicated by an increase in the turbidity of the media in the tube. In this activity we will consider the growth of microbes that aren't measured by turbidity in a test tube, but which are visible to the naked eye–the lowly mold. Using either computer software or manual methods, you will measure the growth of mold on a single slice of bread over the course of ten days. You will then investigate several standard models of growth and consider what these models indicate about the growth of the mold in your cultures.

Overview of the data collection procedures

A full description of the procedures for this lab, including mold plate preparation, image collection, image transfer, image measurement, and image analysis, is available in the "Mold Lab Procedures" section of the Appendix and on the web site.

Establishing the mold cultures

Begin by setting up two series of cultures: *Aspergillus niger* and *Penicillium chrysogenum*. Different species of mold may be used; we have used *Aspergillus niger, Mucor hiemalis* (+) and (-), *Penicillium chrysogenum, Penicillium digitatum, Penicillium italicum, Rhizopus stolonifer* (+) and (-), and *Sordaria fimicola*. If you do not have the facilities to grow and photograph your own mold, photographic datasets of *A. niger, P. chrysogenum*, and *M. hiemalis* (+) are provided on the *Microbes Count!* web site at http://bioquest.org/microbescount

Collecting your data

Once you have established your mold cultures, you will need to record the growth of the mold. To do this you will photograph the cultures at set intervals over a series of days, beginning immediately after you inoculate the bread with the mold. A partial data set of images from *Mucor hiemalis* is shown in Figure 1.

Next, use the photographs of the mold culture to measure the area of mold growth on the bread slices. You can estimate the area of growth in one of two ways. The preferred way is by quantitative image analysis using a freeware tool such as *NIH*

Image (Macintosh) or *Scion Image* (PC). Alternatively, you can use an overlay of transparent graph paper with a manual counting method. You will need to consider how much repetition and averaging you want to use in your data collection. Instructions for both the image analysis and the manual counting methods are in the Appendix and on the web site.

Models of growth

After you have collected your mold growth data, you can use the Excel worksheet file called "Mold Growth Models" to investigate three different models of growth. This worksheet models growth of a mold colony based on user-entered parameter values and models of linear, exponential, and logistic growth. The program will compare the predictions of these growth models with actual data and plot the results graphically.

Introduction to the models

Growth is a key property of many systems: an economic expansion, the formation of a crystal, a religion's increasing congregation, an adolescent's growth spurt, a disease epidemic, and the condensation of a stellar mass. All of these are instances of the same basic process. As a result, we can explore the general properties of these very different events using just three simple mathematical models.

Linear growth

The simplest growth model describes *linear growth*, in which something grows at a constant rate over time. For example, if each day you put two soda cans in a recycling bin, the number of cans in the bin will grow linearly. Linear growth is described by the equation

$$N(t + 1) = N(t) + D.$$

where $N(t)$ represents the numbers or size of the system at time t, $N(t + 1)$ represents the system's numbers or size of the system one time unit later, and D is the system's (linear) growth rate. See Figure 2a.

Figure 2. Examples of different growth models, comparing the experimental data from *Penicillium chrysogenum* to the theoretical curve for each type of growth. These examples are taken from the Mold Growth Models worksheet.

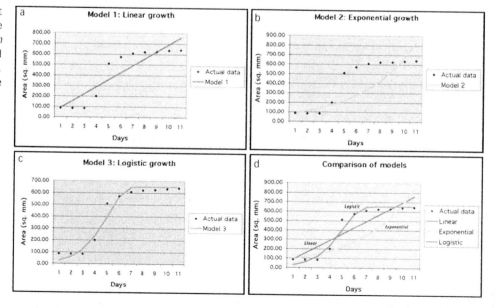

Exponential growth

Another simple model describes *exponential growth*, in which something grows at a constant *proportional* rate over time. For example, if your savings account carries a 5% interest rate, your balance will grow exponentially. Exponential growth is described by the equation

$$N(t+1) = N(t) + r \cdot N(t),$$

where r is the system's (exponential) growth rate. Note that both linear and exponential growth imply continual expansion: either would take the system to infinite size if continued indefinitely. In practice, therefore, each of these growth patterns generally lasts only for a limited time. See Figure 2b.

Logistic growth

A third model, describing *logistic growth*, is similar to exponential growth except that it assumes an intrinsic sustainable maximum, sometimes called the *carrying capacity*. A system far below its carrying capacity will at first grow almost exponentially; however, this growth gradually slows as the system expands, finally bringing it to a halt precisely at the carrying capacity. More complex behaviors can result if the system's growth rate is high (May, 1976). A system starting above its carrying capacity, by contrast, will decrease ever more slowly until it again reaches carrying capacity. This behavior is modeled by the equation

$$N(t+1) = N(t) + \frac{r \bullet N(t) \bullet [K - N(t)]}{K},$$

where K is the system's carrying capacity. Most biological growth processes (as well as many non-biological ones) grow logistically. See Figure 2c.

Using the "Mold Growth Models" worksheet

Open the "Mold Growth Models" file, located in the Modeling Mold folder on at *bioquest.org/microbescount*.

The red-bordered cells under each growth model contain values for the model's parameters. You can replace these default values with other values and observe the effect on the models plotted below. Note that models 1 and 2 have only two parameters, while model 3 has a third (carrying capacity).

Enter your mold growth data, or the data from an existing experiment, in column B, under the "Actual Data" heading. See Figure 3 on the next page.

The worksheet uses the equations above to calculate the area of the mold colony in square millimeters given the parameter values you have chosen. Under each model, it then calculates the summed squared deviations of that model's predictions from the actual data; the smaller this value, the better that model fits the data. Finally, each model is graphically plotted against the data, first individually and then in a combined plot. See Figure 2d.

Figure 3. A screenshot of the Mold Growth Models spreadsheet. The parameters and data can be changed to compare the predictions of the models to the experimental data.

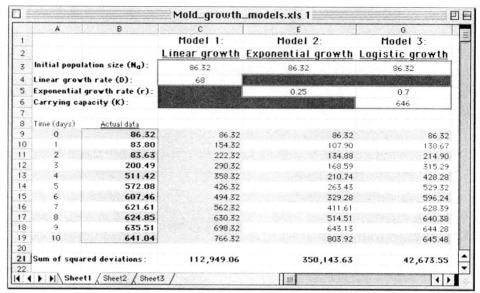

Points to note:

• The sum of squared deviations describes how well each model fits the data *under the assumption* of the particular parameter values entered. Highly inaccurate parameter values will make even the correct growth model seem a poor fit to the data. Since parameters are user-entered, there is no guarantee that they represent a "best fit" for that model. So be sure you can justify the parameter values that you have chosen.

• Model 3 (logistic growth) is essentially a special case of model 2 (exponential growth) with one extra parameter (carrying capacity). Adding parameters to any model will always improve that model's fit to the data unless the fit is already perfect: it's therefore important to determine if the improvement in fit justifies the more complicated model. Standard statistical procedures such as the chi-square test can be applied to answer this question, but they lie outside the scope of this exercise.

Using the "Mold Growth Model" worksheet, consider your growth data under three different sets of assumptions:

1. Growth is independent of the population size that already exists; or

2. Growth is proportional to the population size that already exists; or

3. Growth is both

   a) proportional to the population size that already exists and

   b) the ratio of the population size to the growth rate of the bread mold (which decreases as the population increases).

Consider how these assumptions are related to the following questions:

• Is the growth linear? Exponential? Logistic?

• How could you tell?

- Could you reject one or two models? What additional data would you need to collect?

- How do the two genera differ in their growth behavior? Additional data sets are provided on the *Microbes Count!* web site; you may want to investigate different patterns of growth.

Conclusion

We often use colloquial language to describe quantitative behaviors. As we observed at the beginning of this activity, many microbiologists refer to the growth of bacterial or viral or fungal populations as exponential growth. However, as you've worked with the growth models you have seen that it may be difficult to know if they mean the early phase of logistic growth (pre-leveling off) or if they truly intend an exponential model. We hope that this modeling exercise has helped introduce you to: (a) precision of language, (b) experience in building model complexity by sequentially adding more assumptions or constraints, and, (c) the conversion of verbal, numerical, and pictorial descriptions into symbolic expressions that can be evaluated analytically. We encourage you to repeat these observations or to pool data with classmates in order to improve your ability to accept that the data's fit to the model you have chosen is not due to random effects. What statistical tests can you use? Can you translate your experience in model building here to other laboratories that you have performed?

Software Used in this Activity

NIH Image

Developed by Wayne Rasband at the U.S. National Institutes of Health and available for download at: http://rsb.info.nih.gov/nih-image/

Platform Compatibility: Macintosh only

Scion Image

Developed by Scion Corporation and available for download at: http://www.scioncorp.com

Platform Compatibility: Windows only

Microsoft *Excel*®

Platform Compatibility: Macintosh and Windows

Additional Resources

Available on the *Microbes Count!* web site at http://bioquest.org/microbescount

Text and Data

A PDF copy of this activity, formatted for printing

The "Mold Growth Models.xls" *Excel* model file

"Mold Lab Procedures" document

Photographic data sets for three different kinds of mold

John R. Jungck, Tia Johnson, Anton E. Weisstein, and Joshua Tusin   The BioQUEST Curriculum Consortium

Related *Microbes Count!* Activities

Chapter 2: Shaped to Survive

Chapter 3. Modeling Microbial Growth

Chapter 7: Valuing Variegated Variation: Using Natural Experiments to Understand Viral Infection

Chapter 8: Exploring Microbial Fermentation with Korean Kimchee

Unseen Life on Earth Telecourse

Coordinates with Video I: The Microbial Universe

Relevant Textbook Keywords

Expontential Growth, Growth, Growth curve, Growth Models, Linear Growth, Logistic Growth, Mold, Mold Spores

Related Web Sites (accessed on 4/20/03)

Dresback, M. K. *Fun with Logs*
http://www.unf.edu/~tbratina/kylelog.htm

Microbes Count! Website
http://bioquest.org/microbescount

Unseen Life on Earth: A Telecourse
http://www.microbeworld.org/htm/mam/is_telecourse.htm

References

Kistler, R. A. (1995). Image acquisition, processing & analysis in the biology laboratory. *American Biology Teacher* 57 (3):151-157.

Rasband, Wayne. *About NIH Image*. U.S. National Institutes of Health
http://rsb.info.nih.gov/nih-image/about.html

Bibliography

Blanchard, P., R. L. Devaney, and G. R. Hall (1998). Growth of a population of mold. In *Differential Equations*. Brooks/Cole Publishing Company: California.
http://math.bu.edu/odes (web site with supplemental materials)

May, R. M. (1976). Simple mathematical models with very complicated dynamics. *Nature* 261:459-467.

Robie, J. *The Exciting Calculus of Bread Mold*. On the web at:
http://isolatium.uhh.hawaii.edu/m206L/student%20projects/mold.htm

Schank, J. and W. Wimsatt (2001). Modeling Tools: Logistic Growth. In *The BioQUEST Library Volume VI*. Jungck, J. R. and V. G. Vaughan, Editors. Academic Press: San Diego, CA.

Figure and Table References

Figure 1.   Tia Johnson and Joanna Cramer

Figure 2.   Anton E. Weisstein

Figure 3.   Anton E. Weisstein

Population Explosion: Modeling Phage Growth

Jean Douthwright, Frank Percival, Julyet Aksiyote Benbasat, Tia Johnson, and John R. Jungck

Video I: The Microbial Universe

Bacteria never die, they just phage away - Mark Mueller

Bacteriophage (phage) are viruses that infect bacteria. Like other viruses, they are composed of genetic material surrounded by a protein coat. They can be isolated from sewage, from feces, from springs, and from soil. Bacteriophage are highly specific in the type of bacteria that they can infect. This specificity makes them useful in the laboratory and of interest to medical researchers as a potential alternative to antibiotics to control pathogenic bacteria.

Once a bacterial cell is infected with one or more bacteriophage, the phage reproduce using the machinery of the cells that they infect (the host cells). Several rounds of viral replication may occur before the infection culminates with the bacteria breaking open (lysing), sending thousands of new bacteriophage into the immediate environment. These progeny phage then move on to infect other bacterial hosts. Thus, unlike the continuous growth observed in mold growing on bread or bacteria in culture, viral numbers increase in pulses–latent periods followed by bursts when the population size dramatically increases.

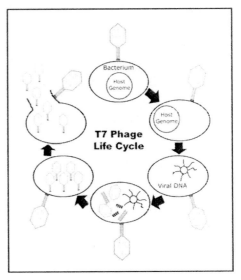

Figure 1. The T7 phage life cycle.

Because bacteriophage can only reproduce within appropriate bacterial hosts, their rate of increase is tied to the growth of the bacteria. In order to model the increase in phage populations, therefore, you must also study the growth of the bacteria they infect. An understanding of the interdependence of these species is critical to developing an accurate model of population growth. In this activity you will model the way particular viral communities (bacteriophage) interact with their host species (bacteria) within their shared ecological community, using a mathematical model of dependent growth.

The interdependence of phage on host bacteria is one of the characteristics that make them attractive as a medical therapy for bacterial disease in humans and other multi-celled organisms. Researchers have found that phage populations in infected individuals expand rapidly, infecting and then killing all available hosts, but that phage numbers crash as they exhaust their supply of hosts. Because these phage are specific to certain strains of bacteria, there is little risk of wiping out helpful bacteria in the host, and less chance of resistance spreading to different species of bacteria. When all of the disease-causing bacteria have been killed, their phage die as well. (See "TB and Antibiotic Resistance" in Chapter 4 for an activity related to the development of antibiotic resistance.) Phage have been used as medical therapy in Eastern Europe for several decades, and research in their use is growing in the United States.

The Phage Growth Spreadsheet Model

Titer: A measure of the concentration of a substance in solution. So a high titer lysate is a solution with a high concentration of phage, measured after the bacteria have burst.

To create a preparation of bacteriophage in the lab, you could mix a small amount of phage with the right bacterial host. When the cells lyse, the solution – now called a lysate – will contain many virus particles. In this process you typically are interested in producing as many virus progeny as possible – that is, you would like to produce a high titer lysate. This is the starting point for producing more concentrated and purified bacteriophage stocks for work in the laboratory.

Discussions of bacteriophage growth typically focus on a one-step growth curve because it provides a framework for thinking about the basic steps of a virus life cycle. When growing phage in the laboratory, however, bacteria are infected with phage, and several rounds of viral replication may occur before the majority of bacteria in the culture are broken open, releasing many phage progeny and producing a high titer lysate (see Figure 1 above.)

In this activity, you will use a model of phage growth to explore some of the factors that affect phage production and the final phage concentration in a high titer lysate. There are six input parameters for the model, all of which you can vary. These parameters are described in Table 1 below.

Table 1: The Input Parameters for the Phage Growth model.

Six Input Parameters		
Parameter Description	Character	Example Value
Initial concentration of uninfected bacteria	B_0	2.00E+07 cells/ml
Bacterial growth rate	Bacterial Doubling Time	30 minutes
Maximum population size of the bacteria, the carrying capacity	Maximum Bacterial Population Size, Bact K	2.00E+09 cells
Initial concentration of phage, initial phage population	P_0	5.00E+03 phage/ml
Number of phage produced per infected bacterial cell, phage growth rate	Burst Size	200 progeny phage/cell
Maximum number of phage which can be attached to a bacterial cell	Phage Binding Sites per Bacterium	250 phage/cell

In this activity you will use a Microsoft *Excel*® spreadsheet to manipulate the model parameters and graph your results. The results of the experiment that you set up when you enter values for the input parameters are calculated using equations already entered into the spreadsheet. These data are displayed in a table and plotted in a graph. To see the model, open the file called "PhageGrowth.xls", located in the Modeling Phage Growth section of the *Microbes Count!* web site.

The first row in each column of the data table contains a short description of how the values in the column are calculated. For additional information, move your cursor over the heading of the column. A box describing the calculation will pop up. You should study these explanations carefully so that you will understand how this simulation models the growth of the bacteriophage population. You will find it much easier to appreciate the complexity of the interactions between the phage and their bacterial hosts if you spend some time working through these calculations. (See the section at the end of this activity for more information on using a spreadsheet model.)

This model of phage growth was developed for use in conjunction with laboratory work involving the production of high-titer phage lysates, but even without that context, there are a host of questions that you can explore with using the spreadsheet model.

Description of the phage growth model

All models are oversimplifications of the phenomena that they are used to represent. Simple models provide two enormous advantages: first, parsimonious models can often account for the major patterns and most of the observed variation in actual experiments; second, if the predictions of the models fail, they provide excellent templates for heuristically investigating what went wrong and why.

Our phage model includes the following assumptions:

- Free phage cannot reproduce; however, they can persist indefinitely.

- The bacteria grow logistically (see "Modeling More Mold" in Chapter 1 and "Modeling Microbial Growth" in Chapter 3 to explore the difference between linear, exponential, and logistic growth). That is, they cannot reproduce at high rates forever because they run out of nutrients and hence reach a carrying capacity in their ecosystem. Thus, their populations eventually reach an equilibrium of births and deaths. In our model, we have an additional cause of death other than starvation–lysis due to phage infection. Furthermore, we assume that infected bacteria do not reproduce.

- We assume that the phage bind randomly to a finite number of discrete receptors on the surface of their host bacteria and that the number of phage bound to a bacterium at one time (the MOI, or multiplicity of infection) is a function of both the number of free phage and the number of currently uninfected bacteria. We don't count phage that are inside of bacterial cells and are not yet released. We assume that all phage that can infect bacteria do so: free phage can exist therefore only when all phage-binding sites of all remaining bacteria are occupied.

- There is simultaneous lysis of infected cells and adsorption of phage to new host cells.

- There is a set time of 20 minutes for phage replication. All infected bacteria lyse synchronously at 20-minute intervals, even if they became infected just one minute ago.

Our model is a null model. Basically, this means that we assume that in the absence of specific knowledge of how multiple phage would bind to a bacterium, separate phage behave independently of one another. That is, no synergism or antagonism between phage is assumed.

Jean Douthwright, Frank Percival, Julyet A. Benbasat, Tia Johnson, & John R. Jungck   The BioQUEST Curriculum Consortium

- In each time interval, the binding of phage to bacteria are independent events.

- The number of phage bound is an integer. That is, we do not count fractions of phage nor do we count partial binding; phage must be fully bound to be infective.

- The probability that a phage binds is small and is proportional to the number of phage, phage receptor sites, and bacteria.

- The probability that more than one phage binds to a single bacterium in a given interval is small compared to the chance of just one binding. (This is similar to a model of radioactive decay and is called an exponential decay function.)

An easy way to visualize the null model is to imagine taking a bag of marbles and dropping them over a rectangular array of boxes – say one box on each square of a chess board. The probability of any box getting more than one marble in any one time interval is proportional to the number of marbles in your bag.

The best way to describe this model is by using a Poisson distribution to represent the infection, growth, and lysis. This function gives the proportion of bacterial cells that are infected with exactly x cells:

where:

- (λt) is the probability that one phage binds to a bacterium;

$$P(x) = \frac{e^{-\lambda t}(\lambda t)^x}{x!}$$

- t is the time elapsed; and

- x is the number of phage bound

We estimate lambda (λ) by using the zeroth class:

One of the nice aspects of using an exponential model such as the Poisson distribution is that any number raised to the zeroth power is equal to one; therefore, we are able to make a good estimate of the value of the coefficients in our model by simply looking at how many boxes on our chess board have no marbles or by observing how many bacteria are uninfected (which we presume also means that they have no phage attached).

For the rest of the model, we are simply tallying who is born and who dies. The three populations that we are concerned with–total number of bacteria, number of infected bacteria, and number of free phage–are plotted as a function of time (Figure 2.)

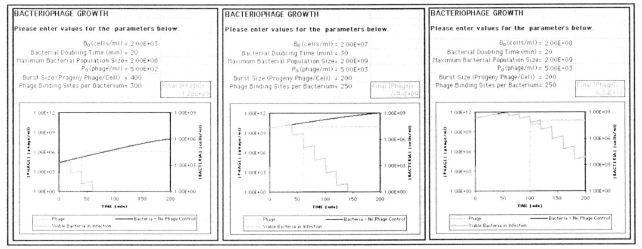

Figure 2. Examples of the three generated values—total bacteria, infected bacteria, and free phage—illustrating how all three interact as a function of burst size and doubling time.

Questions/Exercises

1. Consider a bacterium as a rectangular box with dimensions of 1µm x 1µm x 2µm and the dimensions of a rectangular T7 phage as 60nm x 60nm x 80nm (as given by *The Encyclopedia of Virology*).

   What would be the required viral population to occupy the entire volume of the bacterium, i.e., what is the maximum burst size determined by geometry?

2. Examine the table below:

Time (minutes)	Bacteria (per ml)	Phage (per ml)
0	2.0×10^3	5.0×10^2
10	---	---
20	4.0×10^3	1.77×10^5
30	---	---
40	lysis	---

(Based on *Molecular Biology and Biochemistry: Problems and Applications* by David Freifelder, 1978).

   a. What is the starting concentration of bacteria, B_0?

   b. What is the starting concentration of the phage, P_0?

   c. What is the multiplicity of infection (MOI)? MOI = P_0/B_0

   d. What is the bacterial doubling or generation time? The phage generation time is 20 minutes.

   e. What is the burst size of the phage (the number of phage produced in each host cell)?

   f. Why in this model do all of the bacteria lyse when they do?

   g. How many phage do you expect at lysis?

Jean Douthwright, Frank Percival, Julyet A. Benbasat, Tia Johnson, & John R. Jungck   The BioQUEST Curriculum Consortium

3. You are given:
   - a bacterial stock culture that is 5×10^8 cells/ml, and
   - a phage stock that is 1×10^9 phage/ml

   a. How would you set up the experiment as in the table from Problem 2, using a 10 ml total volume?

   b. Sketch a graph plotting the bacterial and phage concentrations as a function of time. What assumptions have you made?

4. Open the *Excel* spreadsheet "PhageGrowth.xls" and enter B_0, P_0, the bacterial doubling time, and the burst size of the phage, into the proper cells of the spreadsheet. Push the enter key to observe the graph.

   Compare and contrast this new graph with the graph you sketched in question 3.

5. Sketch a graph of infected vs. uninfected bacterial populations, and a graph plotting the bacterial population as a function of phage concentrations. What assumptions have you made?

   This graph is called a phase portrait or plot of your data. A phase portrait plots two population sizes as the X and Y coordinates of each point on the graph; thus, you need to remember that time is a hidden variable and would be represented along the z-axis. It may help you to understand what is going on if you add an arrow along a path or multiple arrows over different portions of a graph to indicate the direction of time.

   a. What did you observe?

   b. How do you relate this to the biological mechanisms involved here?

   c. How are bacteria and phage populations similar to or different from foxes and rabbits as predators and prey in terms of their dynamics?

6. Analyze the phase portraits in Figure 3 below. (See Question 5 for a discussion of phase protraits.)

   a. Why are the phage populations jumping discretely?

   b. Why do the phage populations asymptotically level off?

   c. In the other phage graph, Phage vs Viable Bacteria, why is just the opposite occurring?

   d. If you only looked at the two bacterial populations with respect to one another, what might you infer about the mechanism of the mysterious infective agent?

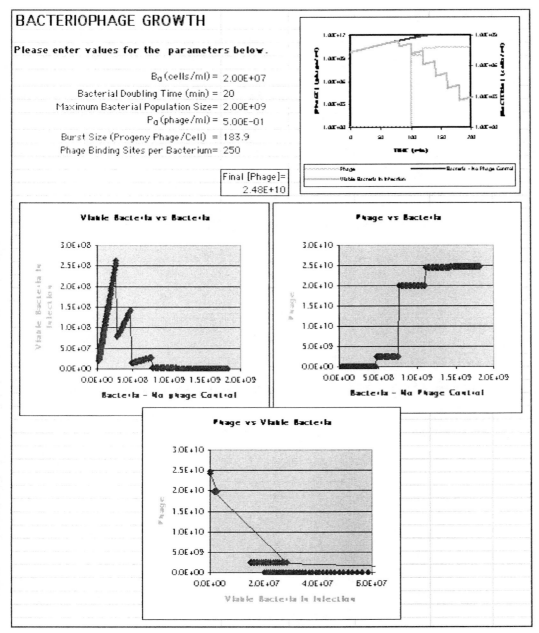

Figure 3. Three phase portraits demonstrating the relationship between two populations, independent of time.

Using the Phage Growth spreadsheet model

To begin your explorations, open the Microsoft *Excel*® file called "Phage Growth.xls". (This file is included in the Modeling Phage Growth section of the Microbes Count web site.) You should see a window containing a graph and some additional information. If you do not see a graph, click on the "Input and Graph" tab at the bottom of the *Excel* window.

The Input and Graph window

This window is where you will set the initial conditions for your experiments and see the results displayed in graphical format. The input parameters are displayed in the top left of the window. To enter a new value, click on the cell containing the current value for the parameter. The contents of that cell will be displayed in the bar below the *Excel* tool bar at the top of the screen. Type in a new value and press Enter. The graph will change to reflect the new value.

The Data Table window

To see the data table for your experiment, click on the "Data Table" tab at the bottom of the window. The table in this window contains the data generated by the experiment that you set up when you set the input parameters in the previous window. In the course of the simulation the values for each time interval are calculated using the values from the previous interval and then plotted in the graph. The first row in each column contains a short description of how the values in the column are calculated. For additional information, move your cursor over the heading of the column. A box describing the calculation will pop up.

Software Used in this Activity

Microsoft *Excel*®
Platform Compatibility: Macintosh and Windows

Additional Resources

Available on the *Microbes Count!* web site at http://bioquest.org/microbescount

Text

A PDF copy of this activity, formatted for printing

The "PhageGrowth.xls" model file

Related *Microbes Count!* Activities

Chapter 1. Modeling More Mold

Chapter 3. Modeling Microbial Growth

Chapter 4. TB and Antibiotic Resistance

Unseen Life on Earth Telecourse

Coordinates with Video I: The Microbial Universe

Relevant Textbook Keywords

Bacteriophage, Growth rate, High titer lysate, Lysis

Related Web Sites (accessed 3/21/03)

American Society for Microbiology
http://www.asmusa.org

Bacteriophages in the news:
http://www.niaid.nih.gov/newsroom/releases/deadlystrep.htm

Microbes Count! Website
http://bioquest.org/microbescount

Phage applications
http://www.ars.usda.gov/is/pr/2001/010921.htm

Unseen Life on Earth: A Telecourse
http://www.microbeworld.org/htm/mam/is_telecourse.htm

Bibliography

Abedon, S. T., T. D. Herschler, and D. Stopar. (2001). Bacteriophage latent-period evolution as a response to resource availability. *Applied and Environmental Microbiology* 67:4233–4241.

Rabinovitch, A., H. Hadas, M. Einav, Z. Melamed, and A. Zaritsky. (1999). Model for bacteriophage T4 development in *Escherichia coli. Journal of Bacteriology* 181:1677–1683.

Schnaitman, C. A. (2002). Phage biology: coming of age. *Science* 298:2329.

Stansfield, W. D. (1991). *Schaum's Outlines Genetics*, 3rd ed. McGraw-Hill: New York. p. 161.

Stone, R. (2002). Stalin's forgotten cure. *Science* 298:728.

References

Freifelder, D. (1978). *Molecular Biology and Biochemistry Problems and Applications*. W. H. Freeman and Company: San Francisco, CA.

Webster, R. G. and A. Granoff, editors (1994). *Encyclopedia of Virology, Vol. 3*. Academic Press, Harcourt Brace & Company, Publishers: London.

Figure and Table References

Figure 1.   Courtesy of Joshua Tusin

Figure 2.   Screenshot from *Excel*

Figure 3.   Screenshot from *Excel*

Chapter 2

Activities for Video II: The Unity of Living Systems

New developments in molecular biology and in genetics offer convincing evidence for metabolic similarities between all living organisms. Traditionally, biologists have dwelt on differences to classify organisms and to explore their relationships to each other. In this chapter we have chosen to emphasize the unity of life. Yeasts and bacteria co-exist within an acidic microenvironment by virtue of shared metabolic products and tolerances. In order to compare microbes with other organisms we need to develop a methodology for determining relative size across the microbial-human continuum. There are surprising similarities in proteins produced by divergent organisms in the break down of starch. None the less we should also consider the metabolic flexibility that a single species exhibits with radical changes in its growth patterns.

In this unit, you can:

- identify the metabolically managed relationships between microbial populations in a sourdough bread culture,
- make microbial measurements that correspond to our own scale,
- investigate the use of bioinformatics to identify potentially useful microbial amylases for industry, and
- observe how nutrient limitation alters the metabolic circumstances resulting in identifiable patterns of growth within the same species.

Sourdough Symbiosis

Who can forget the tangy taste of sourdough? Like other fermented breads, sourdough requires living yeast to make it rise. A sourdough starter culture includes bacteria as well. Can you build a conceptual model to show the relationships between these microbes? How does the model change for the engineered environment for fermentation in corn "wet milling"?

The Scale of the Microbial World

How can we establish a scale for bacteria and viruses that is meaningful? An appreciation of scale is essential for understanding how microbes function– whether we are interested in pathogenesis or in soil enrichment. A data set of microbes and a few other objects with size information is provided for determining scale bars and assigning magnification values.

Searching for Amylase

Information about sequence, structure, and active sites for many proteins is accessible online. This real world application of bioinformatics introduces us to amylases and asks us to explore functionally similar enzymes from markedly divergent organisms. Here, we search for microbial enzymes useful in the starch processing industry. Several tools useful for dealing with sequence information are introduced through the Biology Workbench.

Shaped to Survive

Culturing bacteria under conditions of stress produces colonies with strikingly distinctive shapes that reflect strategies for survival. This fractal growth can be characterized by mapping the borders of the colonies. This exercise provides images of stressed colonies that you can analyze manually or with the program, Fractal Dimension. (You can inoculate your own plates, but be prepared to allow six months for strong patterns to emerge.)

The BioQUEST Curriculum Consortium

Sourdough Symbiosis
Ethel D. Stanley

Video II: The Unity of Living Systems

The role of microorganisms in the biology of "rising" breads remained a mystery until the 1800's, yet the Egyptians were making fermented breads as early as 2600 BC. Archaeologists have discovered murals depicting the baking of bread as well as loaves of bread buried with the dead. Wild yeasts living on the grains used to make flour were an inadvertent, but essential part of the early bread-making process.

The use of yeast that is living and actively metabolizing is critical to the success of breads. Making bread at home sometimes results in small or flat loaves with a dense texture. The most likely cause is a problem with the yeast population. This could be a result of using a packet of dry yeast that is beyond its expiration date so no viable yeast are left to reproduce. The process of reactivating the yeast can also kill the individual cells if the temperature of the added liquids is too high. The optimum temperature range is between 104°F and 114°F. Interestingly, when the temperatures are too low, yeast cells are more likely to lyse than reproduce. This chemically reverses the natural cross-linking of flour proteins known as gluten.

Figure 1. A loaf of sourdough bread, the result of the complex interaction of several kinds of microorganisms.

Bread rises when reproducing yeasts in bread dough produce carbon dioxide that gets trapped within the gluten structure of moistened and kneaded flour. The yeasts use the sugars, including maltose, that are naturally present in bread flour. The maltose is broken down by beta-amylases supplied by sprouted wheat mixed in with the flour. The species of yeast used gives breads their characteristic flavors.

Sourdough Symbiosis

Not all breads share the same microbiology. According to legend, gold miners in California and Alaska couldn't prevent baker's yeast from spoiling. To make their breads rise, they kept a starter of uncooked dough to mix in with their next batch. The resulting breads were more tangy and chewy than regular bread, and the starter didn't spoil from contamination with other microbes.

The mystery of sourdough bread's special characteristics can be understood by studying the symbiosis between the yeast and bacteria present in the starter.

The yeast in sourdough, *Candida milleri* (formerly known as *Saccharomyces exiguus*), is an acidophile that grows best at pH 3.8-4.5. Regular baker's yeast (*Saccharomyces cerevisiae*) prefers a pH of 5.5. Unlike baker's yeast, *Candida milleri* can't metabolize the maltose present in the dough. However, sourdough starter also contains two species of lactobacilli that metabolize available maltose, producing acetic acid and lactic acid. In fact, these lactobacilli can't survive without maltose. The lactobacilli also produce glucose, which is then available to supplement yeast growth.

Candida milleri and *Lactobacillus sanfrancisco* occur in sourdough bread starter in a ratio of 1:100. Few other contaminating microbes are found because lactobacilli secrete an antibiotic called cycloheximide. *Candida milleri*, but not *Saccharomyces cervisiae*, are resistant to the effects of this antibiotic.

Of course these organisms are killed when bread is baked. Saving a small bit of uncooked sourdough and adding flour and water to feed it produces a starter for next week's loaf.

A number of *Microbes Count!* activities ask you to use modeling and simulation software to explore the topic of the lab. Scientists often develop models to help them better understand what is known about relationships in a complex system. The first step in the modeling process involves organizing the information and identifying relationships. A surprising number of questions can arise as a result.

- Make a diagram of the relationships between the microbial populations in the sourdough starter. Be sure to include the roles of maltose, glucose, pH, and cycloheximide.

A fermentation environment similar to that of the sourdough starter is used in the food processing industry for the wet milling of corn. In this case, liquid is added to corn to begin a 1-2 day "steeping" process. Initially a small amount of sulfur dioxide is bubbled into the water producing bisulfite anions which buffer the solution and maintain a pH level of 4. The corn kernels rehydrate and soften. The gluten disulfide bonds (cross-linking of proteins within the corn) are cleaved by bisulfite and facilitate the release of starch.

This steeping environment also affects the microbes present on the corn. While the majority of bacteria are killed, the lactobacilli thrive in this environment. They utilize the sucrose present in the corn. The levels of lactic acid rise sharply in this short period protecting the prepared corn from further microbial contamination. The steeped corn is then ground and separated into its component parts.

Make a second diagram for the steeping process above. Be sure to include the roles of sulfur dioxide, bisulfites, pH and sucrose.

- Use your diagrams to compare these two fermentation processes. Be sure to consider the roles of the lactobacilli.

- What would be different in these two processes without the intervention of humans?

Optional Activity

1. The table below lists four fermented products you could find in local stores, as well as a plant and a microbe necessary to make the product. Add four more fermented products that are less likely to be found in local stores.

Country	Product	Plant	Microbe
United States	Cocoa	Cacao	*Candida krusei*
United States	Coffee	Coffee	*Erwinia dissolvens*
United States	Pickles	Cucumbers	*Lactobacillus plantarum*
United States	Bourbon	Corn	*Saccharomyces cerevisiae*

2. Not all wild microbes associated with grain crops are beneficial to humans. In Europe, unusually damp weather during harvest season has been historically linked to food poisoning. In areas where rye bread was included as a substantial part of the daily diet, ergot poisoning from the microorganism *Claviceps purpurea* is thought to be responsible.

• Describe how ergot metabolism differs from that of yeast in its effects on human metabolism

Note: You may find it interesting to read more about these outbreaks of ergot poisoning and their effects on local culture.

Additional Resources

Available on the *Microbes Count!* web site http://bioquest.org/microbescount

Text

A PDF copy of this activity, formatted for printing

Related *Microbes Count!* Activities

Chapter 3: Modeling Wine Fermentation

Chapter 8: Exploring Microbial Fermentation with Korean Kimchee

Unseen Life on Earth Telecourse

Coordinates with Video II: The Unity of Living Systems

Relevant Textbook Keywords

Acidophile, *Candida*, Cycloheximide, Metabolism, *Saccharomyces cerevisiae*, Symbiosis

Related Web Sites (accessed on 2/20/03)

Microbes Count! Website
http://bioquest.org/microbescount

To learn more about the history of bread:
http://www.bakersfederation.org.uk/histbread.htm
http://www.breadinfo.com/history.shtml
http://www.history-magazine.com/bread.html
http://www.botham.co.uk/bread/history1.htm
http://www.cyberspaceag.com/breadhistory.html

To learn more about the science of bread:
http://www.exploratorium.edu/cooking/bread/links.html
http://www.faqs.org/faqs/food/sourdough/faq/section-4.html
http://www.landfield.com/faqs/food/sourdough/faq/section-21.html
http://www.redstaryeast.net/science.htm
http://members.tripod.com/~FungiFood/breadbio.htm

To learn more about the wet milling of corn:
http://www.corn.org/web/tapping.htm
http://www.epa.gov/ttn/chief/ap42/ch09/final/c9s09-7.pdf

Unseen Life on Earth: A Telecourse
http://www.microbeworld.org/htm/mam/is_telecourse.htm

References:

Whistler, R. L., J. N. BeMiller, and E. F. Paschall, Eds. (1984). *Starch: Chemistry and Technology* San Diego: Academic Press.

Figure and Table References

Figure 1.   http://www.countrybaker.com/Photos/Recipes/18.jpg

The Scale of the Microbial World

Marion Field Fass

Video II: The Unity of Living Systems

In a world where you were as tall as the Sears Tower, how big would a single spore of *Bacillus anthracis* be?

We can calculate this if we know the height of the Sears Tower (442 meters), your height (1.676 meters if you're 5 feet 6 inches tall), and the diameter of the *B. anthracis* spore (average size is 4 micrometers or 4×10^{-6} meters). The Sears Tower is 264 times bigger than you are. A spore of *Bacillus anthracis* would therefore be 264 times bigger as well, which is $264 \times (4 \times 10^{-6}$ meters or 1.056×10^{-3} meters–about 1 millimeter in size. Just visible, but pretty small!

To understand microbiology, you must be able to imagine the unseeable and be able to relate the microworld to your own. Great microbiologists learn to "see" their microorganisms and appreciate their complexity and skills at surviving in difficult environments. Beginning microbiologists struggle to visualize bacteria and to come to grips with issues of scale as they learn about the relations between eukaryotic cells, bacteria and viruses and their interactions in the environment and as agents of disease.

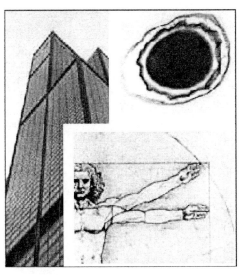

Figure 1. Collage of images including the Sears Tower, spore of *B. anthacis*, and DaVinci's man.

The appreciation of scale is important to an understanding of the processes of pathogenesis. How do antibodies recognize viral particles? How many HIV particles can burst forth from an infected lymphocyte? Is it easy to see a bacterium infecting a cell? If you were asked to make a sketch of streptococcus bacteria attacking a human epithelial cell, how large would the bacteria appear in comparison to the epithelial cell? Just what would an "attack" look like?

Scale is a necessary consideration as microbiologists strive to understand the microenvironments bacteria inhabit. The complex community in a pinch of soil is only comprehensible to scientists who appreciate the scale of the organisms they are working with.

It is challenging to establish an understanding of scale for bacteria and viruses in a meaningful way. Yes, we can discuss how many bacteria will fit on the head of a pin, but we don't really have much of a sense of how a pin compares to the size of our own body cells. You might wish to explore the web resource called *Cells Alive* (www.cellsalive.com) for a great animation that demonstrates the size of microbes. We can look at microbes under the microscope in the lab, but when even the best light microscopes show bacteria only as small dots and lines, it is hard to gauge size. How big is that microscopic dot in the field?

Activity 1. Calculating relative sizes using a human scale

How can we develop a better sense of the relative size of the living entities in the microworld? First, you will need to become familiar with the metric units for length that are commonly used for measurement in the microworld. Note that the units differ by multiples of ten.

Table 1. Metric measurements

1 meter	1×10^{0} meter
1 millimeter	1×10^{-3} meter
1 micrometer	1×10^{-6} meter
1 nanometer	1×10^{-9} meter

Now, to put your measurements in familiar terms, calculate the size of the microbes on a human scale. For example, if an *E. coli* bacterium, actual size about 0.002 millimeters, is magnified to a size equal to your own height, how many times larger than its actual size would it be? That is, what is the factor needed to convert from the actual size to a size relative to a human of your height? Divide your height by the length of the bacterium. This is your conversion factor. So, if you are 5 feet 6 inches (1.676 meters) tall, an *E. coli* as tall as you would be 838,000 times larger than its actual size.

On this human scale, in which an *E. coli* is as big as you are, how big would a red blood cell be? A red blood cell (rbc) is about 10 micrometers in diameter. To calculate the size of the rbc relative to a human, multiply its actual size by the conversion factor you calculated above.

Another way to make this conversion is to use the proportional relationship below. (The calculations assume a height of 1.676 meters; you'll want to use your actual height.)

$$\frac{\textit{Your Height}}{\textit{Actual Bacterium Length}} = \frac{\textit{Relative RBC Size}}{\textit{Actual RBC Size}}$$

$$\frac{\textit{1.676 meters}}{\textit{2 micrometers}} = \frac{x}{\textit{10 micrometers}}$$

Calculate the relative size of each of the items in the following table:

	Actual Size	Size to Scale
An Influenza virus	100 nanometers	
The Smallpox virus	200-350 nanometers	
Bacillus anthracis	1-1.5 by 4-10 micrometers	
A spore of *Bacillus anthracis*	2-6 micrometers	
The largest life form of *Pfeisteria pisicida*	450 micrometers	
The smallest form of *Pfeisteria pisicida*	5 micrometers	
A human red blood cell	10 micrometers	
The width of a strand of human hair	0.1 millimeters	
T 7 bacteriophage	40 nanometers	
Saccharomyces cerevisiae	4 micrometers	
HIV	80-100 nanometers	
Average human epithelial cell	10 micrometers	
Lactobacilli	3-5 micrometers	
Paramecium	200 micrometers	

Activity 2. Comparing sizes under the microscope

A strand of hair provides an easy comparison for the scale of microbes, because a hair is both visible to the eye and able to be studied under the light microscope. It is too big, however, to be seen in its entirety under the 100x lens of the microscope.

If you have access to a light microscope and prepared slides of bacteria and protozoa, look at five different microbes under the microscope. Use the information in the table below, or information provided by your instructor for field sizes (microscopes vary), to determine the actual size of the microbes you see magnified.

Eyepiece Magnification	Objective Magnification	Total Magnification	Viewfield Diameter
10x	4x	40	4.5 mm
10x	10x	100	1.8 mm
10x	40x	400	0.45 mm
10x	100x	1000	0.18 mm

- How do your estimates compare to the sizes identified in your textbook or lab manual?

- How can you explain the discrepancies?

Activity 3. Determining size from scale

Although you may have access to images of microbes in your text, this activity requires you to search the Internet for images of microbes. One useful search strategy is to go to Google.com and select the image option for your search. The *ASM Microbe Library* visual resources include photomicrographs that are excellent for exploring the relative size of microbes. Many of these images indicate the scale or the size of the organism, but some do not.

- Find examples of two bacteria, two viruses and two eukaryotes. Identify the method used to visualize each organism, its actual size, and what it does in the world.

- For some of these organisms, it may be necessary to calculate the size from the magnification given in the description of the image.

Additional Resources

Available on the *Microbes Count!* web site at http://bioquest.org/microbescount

Text

A PDF copy of this activity, formatted for printing

Related *Microbes Count!* Activities

Chapter 1: Population Explosion: Modeling Phage Growth

Chapter 3: The Living World of Yogurt

Unseen Life on Earth Telecourse

Coordinates with Video II: The Unity of Living Systems

Relevant Textbook Keywords

Measurement, Microscope, Microscopy, Size

Related Web Sites (accessed on 2/20/03)

American Society for Microbiology
http://asmusa.org

Image of a Macrophage Cell Attacking Bacteria
http://www.people.virginia.edu/~rjh9u/macro.html

Length Conversion Calculator
http://www.worldwidemetric.com/metcal.htm

Microbes Count! Website
http://bioquest.org/microbescount

Unseen Life on Earth: A Telecourse
http://www.microbeworld.org/htm/mam/is_telecourse.htm

References (web sites accessed on 3/8/03)

Microbe Library Visual Resources
http://www.microbelibrary.org/Visual/page1.htm

Cells Alive
http://www.cellsalive.com

Figure and Table References

Figure 1.   Collage courtesy of Ethel D. Stanley

Searching for Amylase
Keith D. Stanley and Ethel D. Stanley

Video II: The Unity of Living Systems

Food chemists utilize a surprising variety of microbes in the manufacture of processed foods. The next time you snack on chips, cookies, ice cream, or a soda, look at the ingredients listed on the packaging. Chances are you'll find maltodextrin (partially hydrolyzed food starch) and either dextrose or high fructose corn syrup among them.

Although corn starch is used to make each of the ingredients above, the specific process differs. The breakdown of starch must be carefully controlled using both physical and chemical means. Enzymes useful in corn starch processing have been isolated from a number of organisms, many of which are microbes. For example, an alpha-amylase found in *Bacillus licheniformis* is commonly used to make maltodextrins.

© Tate & Lyle

Figure 1. This commercial ice cream product contains both maltodextrin and high fructose corn syrup.

Alpha-amylase breaks down starch by cleaving alpha 1,4 linkages between glucose monomers randomly within the chain. The partially hydrolyzed starches that result are then removed as maltodextrins. All alpha-amylases will do this, but with different efficiencies under different conditions. Most bacterial amylases, for example, are active at temperatures as high as 50 to 70 degrees Celsius. However, the alpha-amylase from *B. lichenformis* is active at 90 degrees Celsius, which is also the temperature necessary to dissolve corn starch in water. Alpha-amylases found in humans lose reactivity at such high temperatures.

One of the most valuable products from starch in today's market is high fructose corn syrup. This ingredient often replaces sugar to enhance flavors or add

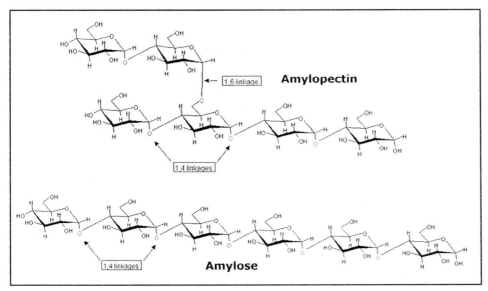

Figure 2. Starch is composed of amylopectin and amylose. The straight chain amylose contains 1,4 linkages between glucose monomers. Amylopectin is branched with 1,6 linkages in addition to 1,4 linkages. (The actual amylose and amylopectin molecules contain thousands of glucose monomers.)

sweetness to products like soda pop. To make high fructose corn syrup, starch is first converted to high dextrose corn syrup by:

1. adding alpha-amylase to the heated starch solution to produce smaller starch fragments,

2. treating with glucoamylase (an amyloglucosidase), which hydrolyzes the starch fragments by cleaving glucose units from the non-reducing end, and

3. in some cases, simultaneously adding a debranching enzyme such as pullulanase that hydrolyzes starch by cleaving alpha 1,6 linkages.

Finally, to produce high fructose corn syrup, a glucose isomerase is used to partially convert the dextrose to fructose.

Microbial enzymes have been examined and found to be effective in other kinds of corn starch processing as well. Table 1 shows some of the microbes that produce enzymes commonly used in industry.

Table 1: Sources for the industrial enzymes used to make high fructose corn syrup.

Microbe	Enzyme
Bacillus licheniformis	Alpha-amylase
Aspergillus niger	Glucoamylase
Klebsiella aerogenes	Pullulanase
Bacillus circulans	Glucose isomerase

- Do any of the organisms in Table 1 make starch themselves?

- Why do you think these organisms produce starch-processing enzymes?

Searching for Industrial Enzymes

What makes one alpha-amylase better than another for specific starch processing tasks? What do chemists look for?

Thermal stability at high temperatures is one desirable characteristic for starch processing. Specificity (how an enzyme interacts with a substrate) is even more important. Enzymes have active sites which include both catalytic sites and binding subsites. While the amino acid sequence of different alpha-amylases may vary, there are specific aspartic acid and glutamic acid units found in the beta-strands of the molecule that are responsible for the catalysis of glycosidic bond cleavage. Other amino acids such as histidine are involved in establishing conformation and binding of the substrate. Binding subsites also affect enzyme efficiency. For example, pig alpha-amylase and barley alpha-amylase have similar catalytic sites, but their binding subsites are different. Pig alpha-amylase has five subsites, while as many as ten subsites are suspected in barley. (MacGregor, 1993)

How do you find new sources of industrial enzymes?

Well, you don't necessarily have to go to the field. Today, accessing data from existing molecular databases and utilizing online bioinformatics and visualization tools are invaluable in tracking down candidates for the next generation of industrial microbes.

You can find sequence information for an enzyme like the TAKA-amylase from *Aspergillus oryzae,* which is used commercially, by searching online molecular databases such as SWISS-PROT. Specifying keywords such as amylase and Aspergillus, an accession number, or a specialized label such as AMYA_ASPOR will allow you to isolate the protein and its molecular information. In the *Biology Workbench*, the search function is called Ndjinn.

(For an overview of the *Biology Workbench* and how it is organized, please see the "Orientation to the *Biology Workbench*" document on the *Microbes Count!* web site. You may also want to take a look at the "Proteins: Historians of Life on Earth" and "Tree of Life: An Introduction to Microbial Phylogeny" activities in Chapter 6 for some examples of using the Biology Workbench.)

In Figure 3, the highlighted sequence GLRIDTVKH is a conserved region that functions as an active site in alpha amylase. See Table 2 for more active site information.

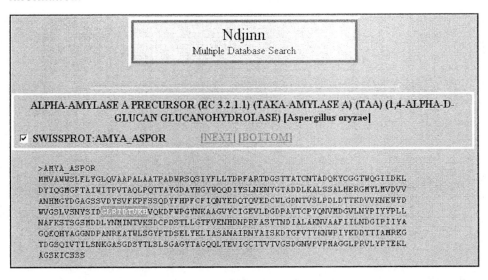

Figure 3. The amino acid sequence for AMYA_ASPOR in the SWISSPROT database was retrieved using Ndjinn on the *Biology Workbench* site.

Table 2: Well-conserved regions of alpha-amylase from *Aspergillus oryzae* with locations of specific amino acids and known structure-function relationships.

Starting and ending position of residues	Conserved sequence (Highly conserved in bold)	Some known structure - function relationships	
116 - 122	VDVVANH	Subsite 1	H = His 122
202 - 210	GLRIDTVKH	Catalytic site Subsite 1 Subsite 2	D = Asp 206 H = His 210 K = Lys 209
230 - 233	EVLD	Catalytic site	E = Glu 230
293 - 298	VENHDN	Catalytic site Subsite 1	D = Asp 297 H = His 296

The AMYA_ASPOR sequence can be used with a BLAST procedure to obtain the records for proteins having similar sequences (Figure 4).

Figure 4. Searching for similar protein sequences using BLASTP on the *Biology Workbench* site.

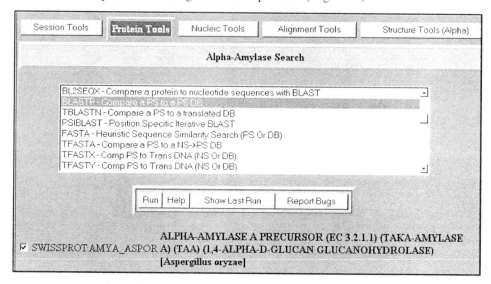

A subset of the sequences for these similar proteins are chosen for further study. The sequences are extracted and compared systematically with the CLUSTALW procedure which aligns the sequences (Figures 5).

Figure 5. Sample output showing a portion of the aligned sequences using CLUSTALW on the *Biology Workbench* site.

```
AMY_BACCI        NNSTIDTYFKNAIR-LWLDMGIDGIRVDAVKHMPFGWQKNWMSSIYSYKPVFTFGEWFLG
CDGT_THETU       QNSTIDSYLKSAIK-VWLDMGIDGIRLDAVKHMPFGWQKNFMDSILSYRPVFTFGEWFLG
AMYR_BACS8       NNSTSDVYLKDAIK-MWLDLGIDGIRMDAVKHMPFGWQKSFMAAVNNYKPVFTFGEWFLG
AMY3_WHEAT       LNPRVQRELSAWLNWLKTDLGFDGWRLDFAKGYSAAMAKIYVD---NSKPAFVVGELY--
AM3C_ORYSA       LNTRVQTELSDWLNWLKSDVGFDGWRLDFAKGYSATVAKTYVD---NTDPSFVVAEIWSN
6056337_6056338  EKDYVRSMIADYLN-KLIDIGVAGFRIDASKHMWPGDIKAVLDKLHNLNTNW-FPAGSRP
397984_397985    GNQWVRDRIVDLMN-KCVGYGVAGFRVDAVKHMWPGDLEHIYSRLNNLNTDHGFPHGAKP
AMYA_ASPOR       TKDVVKNEWYDWVGSLVSNYSIDGLRIDTVKHVQKDFWPGYN----KAAGVYCIGEVLDG
```

This subset of aligned sequences can be used to generate trees that group the proteins according to their degree of similarity (Figure 6). Note: this procedure very broadly defines the relationships since not all sequence data carry the same weight.

Figure 6. Unrooted tree for selected proteins generated by a ClustalW search on the *Biology Workbench* site.

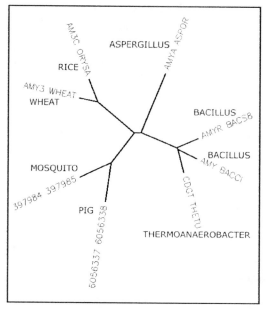

Label and Name of Organism		Conserved Sequence
AMY_BACCI	*Bacillus*	GIRVDAVKH
CDGT_THETU	*Thermanoaerobacter*	GIRLDAVKH
AMYR_BACS8	*Bacillus*	GIRMDAVKH
AMY3_WHEAT	Wheat	GWRLDFAKG
AM3C_ORYSA	Rice	GWRLDFAKG
6506337_605633	Pig	GFRIDASKH
397984_397985	Mosquito	GFRVDAVKH
AMYA_ASPOR	*Aspergillus*	GLRIDTVKH

Table 3. Amino acid sequences that correspond to positions 202-210 in the Aspergillus amylase AMYA_ASPOR are compared for similar proteins found in a selected group of diverse organisms.

Looking at an example with beta-amylases

Beta-amylases are used in starch processing to make high maltose corn syrup which is useful in the brewing industry. Let's use a general bioinformatics approach to look at beta-amylases in several different organisms and make some recommendations for potential use in industry.

Beta-amylases, like all proteins, are made up of sequences of amino acids. The sequences that make up the active sites in the protein tend to be more highly conserved; that is, they do not show as much variation as sequences in the molecule that are not directly involved in enzymatic activity.

Active Site	Conserved Amino Acid Sequence
Site 1	HxCGGNVGD
Site 2	Gx<SA>GE<LIVM>RYPSY

Table 4. Conserved sequences for two of the catalytic sites in beta-amylase.

In Table 4, the amino acid sequence is HxCGGNVGD for active site 1. These one-letter codes represent a different amino acid. For example, H refers to the amino acid histidine. The lowercase x indicates that any one of several different amino acids could be found in that specific position. (See your text book to interpret codes.)

The sequence for active site 2 is Gx<SA>GE<LIVM>RYPSY. Here the letters in brackets, such as <SA>, indicate that any one of the amino acids shown is found in this position.

- Do all of the following sequences fit the sequence algorithm for active site 2?

  GPSGELRYPSY

  GAAGELRYPSY

  GPAGEMRYPSY

Figure 7: Sequences are matched to the regions of active sites in a wire frame model of the beta-amylase produced by the bacterium *Thermanoerobacter thermosulfurogenes.*

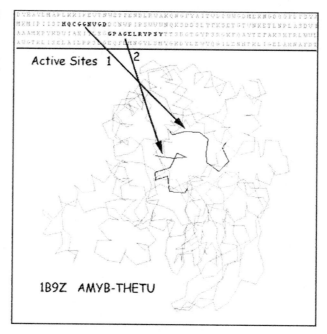

Active Sites 1 2

1B9Z AMYB-THETU

In Figure 7 above, the arrows link two amino acid sequences to their sites in the model of the beta-amylase enzyme. The two catalytic sites appear closer to each other in this model of the enzyme than the amino acid sequences are.

• Suggest a reason for their proximity?

Note: The open area in the center is the substrate binding site.

Scientists can infer something about the expected activity of similar proteins produced in different organisms based on similarities and differences in the active sites. The table below shows beta-amylase sequences from the SWISS-PROT database for two active sites in soybean and the bacterium, *Bacillus circulans.*

• Fill in the sequences for the other organisms by searching SWISS-PROT using the label provided and then aligning sequences with ClustalW.

Label and Name of Organism		Active Site 1	Active Site 2
AMYB_SOYBN	Soybean	HQCGGNVGD	GPAGELRYPSY
AMYB_ARATH	Mouse-Ear Cress		
AMYB_IPOBA	Sweet Potato		
AMYB_HORVU	Barley		
AMYB_PAEPO	*Bacillus polymyxa*		
AMYB_BACCI	*Bacillus circulans*	HRCGGNVGD	GPSGELRYPSY
AMYB_THETU	*Clostridium thermosulfurogenes*		
AMYB_BACCE	*Bacillus cereus*		

- Besides the active sites, what other kinds of information would you like to know about a beta-amylase enzyme before recommending it for use in the starch processing industry?

- AMYB_THETU is a commonly used beta-amylase in corn starch processing. This enzyme is produced by *Thermanoaerobacter thermosulfurogenes (also known as Clostridium thermosulfurogenes)*. Recommend a different beta-amylase for chemists at a corn starch processing plant to work with. Explain why you think it would be a good choice for commercial use.

- Find a beta-amylase you would *not* recommend. Explain.

Web Resources Used in this Activity

Biology Workbench

The *Biology Workbench* was originally developed by the Computational Biology Group at the National Center for Supercomputing Applications at the University of Illinois at Urbana-Champaign. Ongoing development of version 3.2 is occurring at the San Diego Supercomputer Center, at the University of California, San Diego. The development was and is directed by Professor Shankar Subramaniam.

Platform Compatibility: Requires an internet connection and a current web browser.

Additional Resources

Available on the *Microbes Count!* web site at http://bioquest.org/microbescount

Text

A PDF copy of this activity, formatted for printing

"Orientation to the *Biology Workbench* "

Related *Microbes Count!* Activities

Chapter 2: Sourdough Symbiosis

Chapter 4: Molecular Forensics

Chapter 4: Exploring HIV Evolution: An Opportunity for Research

Chapter 6: Proteins: Historians of Life on Earth

Chapter 6: Tree of Life: Introduction to Microbial Phylogeny

Chapter 6: Tracking the West Nile Virus

Chapter 6: One Cell, Three Genomes

Chapter 7: Visualizing Microbial Proteins

Unseen Life on Earth Telecourse

Coordinates with Video II : The Unity of Living Systems

Relevant Textbook Keywords

Amylase, Enzymes, Monomers, Specificity, Substrate

Related Web Sites (accessed on 3/4/03)

Biology Workbench
http://workbench.sdsc.edu

Industrial Starch Processing
http://home3.inet.tele.dk/starch/

Microbes Count! Website
http://bioquest.org/microbescount

Unseen Life on Earth: A Telecourse
http://www.microbeworld.org/htm/mam/is_telecourse.htm

References

Guzman-Maldonado, H. and O. Paredes-Lopez (1995). Amylolytic enzymes and products derived from starch: a review. *Critical Reviews in Food Science and Nutrition. 35 (5): 373–403.*

MacGregor, E. A. (1988). a-Amylase structure and activity. *J. Protein. Chem.* 7:399-415.

MacGregor, E. A. (1993). Relationships between structure and activity in the alpha-amylase family of starch-metabolising enzymes. *Starke* 45:232–237.

MacGregor, E. A. (1996). Structure and activity of some starch metabolizing enzymes in *Enzymes for Carbohydrate Engineering*. Park, K-H., J. F. Robyt, and Y. D. Choi, Editors. Elsevier: Amsterdam.

Stanley, E. and K. Stanley (2001). Looking into the Glycosidases: A Bioinformatics Resource for Biology Students. In *The BioQUEST Library Volume VI*, Jungck, J. R. and V. G. Vaughan, Editors. Academic Press: San Diego, CA.

Figure and Table References

Figure 1.   Photo courtesy of Tate and Lyle (http://www.tateandlyle.com/)

Figure 2.   Courtesy of Keith D. Stanley

Figure 3.   Modified from *Biology Workbench* (http://workbench.sdsc.edu)

Figure 4.   Modified from *Biology Workbench* (http://workbench.sdsc.edu)

Figure 5.   Modified from *Biology Workbench* (http://workbench.sdsc.edu)

Figure 6.   Modified from *Biology Workbench* (http://workbench.sdsc.edu)

Figure 7.   Modified from a CHIME image of 1B9Z generated by QuickPDB.
             http://www.rcsb.org/pdb/cgi/explore.cgi?job=graphics&pdbId=1B9Z&page=&pid=235141041544236
             Click on QUICKPDB

Table 1.   Guzman-Maldonado & Paredes-Lopez, 1995

Table 2.   Adapted from MacGregor, 1996

Table 3.   *Biology Workbench* (http://workbench.sdsc.edu/)

Table 4.   From the Prosite website: Accession number PS00506
             http://ca.expasy.org/cgi-bin/prosite-search-ac?ps00506

Table 5:   *Biology Workbench* (http://workbench.sdsc.edu/)

Shaped to Survive

John R. Jungck, Tia Johnson, and Joshua Tusin

 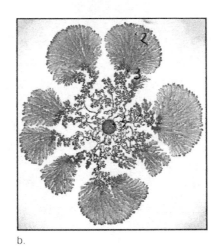 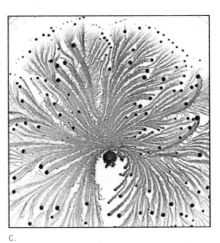

a.   b.   c.

Figure 1: Different types of fractal bacteria colonies. Colonies 1.a and 1.b are *Paenibacillus dendritiformis*; colony 1.c is *Paenibacillus vortex*. These are all colonies that have been grown on petri plates by the Bacterial Cybernetics Group, headed by Eshel Ben-Jacob.

Artistry of phase space of fractal colony morphology for *Bacillus subtilis* grown in different amounts of nutrients and hardness of surface

When bacteria make the cover of magazines that appear on the grocery store shelf, it is usually because we are going to read about another horrific disease that has captured the public's attention. But in October 1998, the cover of *Scientific American* prominently featured a striking fractal image of bacteria growing on a Petri plate. The accompanying article, "The Artistry of Microbes: Shaped to Survive" by Eshel Ben-Jacob and Herbert Levine, appears in a section titled "Science in Pictures." Rarely are art, biology, and mathematics merged in such a beautiful synthesis.

Pattern formation in biological organisms is an area of growing interest, due in part to the advent of chaotic dynamics and the ongoing improvement of computer modeling. Fractal dynamics appear to be present in all sorts of self-organizing systems, from growth patterns of cities to the growth morphology of fungal and bacterial colonies (Ben-Jacob et al. 1994, Fujikawa and Matsushita 1989, 1991; Jones et al. 1993; Makse et al. 1995).

Several species of bacteria have been shown to demonstrate patterns of fractal growth morphology when grown under appropriate conditions. Examples of different levels of fractal growth in *Paenibacillus dendritiformis* are illustrated in Figure 2 on the next page (Bacterial Cybernetics Group, headed by Eshel Ben-Jacob.).

Figure 2: Colonies of *Paenibacillus dendritiformis* grown on hard agar and under increasingly severe starvation. Note that the perimeters of the three different growth levels would have extraordinarily different fractal dimensions (measures of roughness).

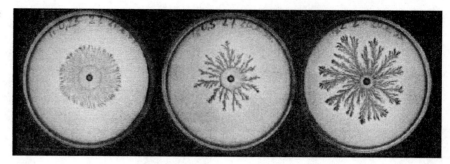

Fujikawa and Matsushita (1989, 1991) have shown that the colonial morphology of *B. subtilis* is very similar to diffusion-limited aggregation (DLA) as a function of nutrient and agar concentrations of an agar plate. (This and other technical terms are explained in the glossary at the end of this activity.) See also Figure 3 from the Bacterial Cybernetics Group. Differences in colonial morphology may reflect strain-specific differences such as growth and mutation rate.

Figure 3: The x-axis denotes the amount of nutrient (peptone) and the y-axis the hardness of the surface (amount of agar).

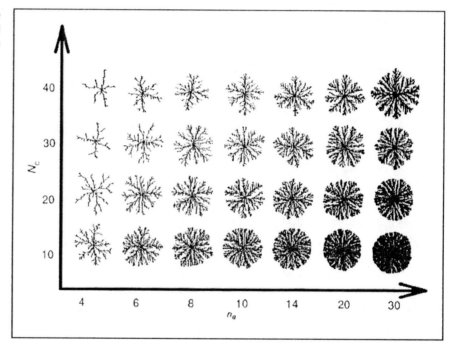

The goal of this activity is to develop and test hypotheses about fractal growth patterns in bacteria.

Consider the following questions:

1. In fungal and bacterial colonies, is fractal morphology a key indicator of environmental conditions, including composition of medium (density of agar, concentration of nutrients and/or vitamins), temperature and time of incubation, and humidity?

2. What can you infer from these patterns about the behavior of bacteria in different environments?

3. How might cell-to-cell communication influence fractal growth?

4. What can you infer about self-organization of biological patterns?

5. How could you extrapolate from the ecology of a petri dish environment, as studied in this activity, to microorganisms growing in harsh environments in natural settings?

For this investigation, you will investigate the patterns of bacterial growth and morphology that result when *Bacillus subtilis* and *Enterobacter aerogenes* are incubated under varying conditions of nutrient and agar concentration. The patterns shown in Figure 3 may be helpful.

To create cultures of *Bacillus subtilis* and *Enterobacter aerogenes* for analysis, consult the directions in Paul Trunfio's (1993) manual on fractal analysis. Be sure to allow adequate time for fractal growth–for our lab we incubated the plates for 16 to 24 weeks. The techniques for digital imaging and analysis are described in the "Analysis Procedures" document available on the *Microbes Count!* web site.

Two references that you might find useful:

1. Hartvigsen (2000) discusses some biological uses of fractals and introduces a method for determining the fractal dimension of a colony by hand, without magnification or digital processing. His technique, also described in the "Box Counting" document on the *Microbes Count!* site, uses a variety of transparencies with different size grids to determine the fractal dimension.

2. "Fractal Geometry" (anon.) is an excellent introduction to applications of fractal analysis in biology. Not only does it cover some basics, but it also makes some important points that even those familiar with the techniques and literature may want to consider. It is available from the *NIH Image* web site that discusses *NIH Image* add-ons.

- Develop a poster presentation using images of fractals analyzed by one of the methods described above. (A detailed description of the analysis procedures is available in the "Analysis Procedures" and "Box Counting" documents on the *Microbes Count!* site.) Identify the question you have posed, the steps you have taken to answer it, your conclusions and questions for further research.

- Consider how you would apply fractal analysis to other areas of microbiology as well. For example, biofilms have been extensively studied with fractal analysis. Unfortunately, most of the research literature uses research-grade equipment like confocal microscopes that many undergraduates do not have access to. How might you employ lower-tech approaches to do similar work? How would you include modeling of DLA and cellular automata to develop a better mathematical understanding of assumptions that we make about growth?

John R. Jungck, Tia Johnson, and Joshua Tusin   The BioQUEST Curriculum Consortium

Procedure for Fractal Analysis

The process of converting a detailed and fairly complicated photograph of a bacterial colony into a simplified image whose fractal dimension can be readily determined is much easier than it might appear at first glance. It can quickly be accomplished using commonly available image processing applications. So that you can get an idea how this conversion is done, the steps in the process are outlined in simplified form below; a more detailed description is available on the *Microbes Count!* web site.

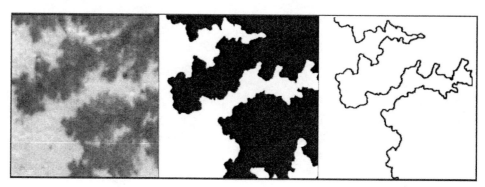

Figure 4. Image preparation. First, the original photo taken of the fractal bacteria is cropped for analysis. Second, the image is altered using the Threshold function in image analysis software. Third, the image is reduced to an edge using the Stroke function. This image is now ready to be taken into the *Fractal Dimension* application for further analysis.

Step 1. Preparing the Image for Analysis

The photograph of the bacteria is transferred to an image editing application (we used *Photoshop*) and the section you wish to analyze is selected and copied to a new image file. Then the following techniques are used to prepare the image for fractal analysis:

a. The section of the colony that you wish to analyze is isolated from the background and converted to a binary (black and white) image.

b. Using the Stroke tool in *Photoshop*, or its equivalent in other applications, the edge of the image is converted to a single line. The transformed image is then saved in a format that can be opened by the *Fractal Dimension* application.

Step 2. Using Fractal Dimension

a. Open the transformed image file in the *Fractal Dimension* application. You will use the "Do Box Measurement" function with boxes of varying size to collect data on the number of boxes it takes to cover the line in

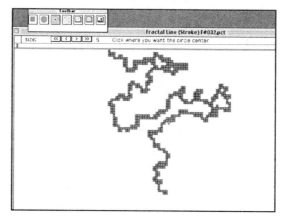

Figure 5. *Fractal Dimension* after the Box Measurement function has been performed. You can see the toolbar and box number arrows in the upper left corner.

your image (Figure 5). The program will automatically add the data for each box size to the "Box Data" window.

b.   When you have collected all of the data that you need, the Graph function will create a graph with a linear fit of the data and a calculated slope (Figure 6).

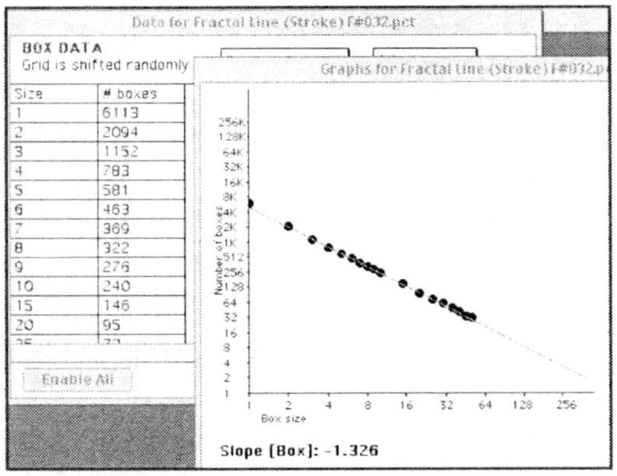

Figure 6. Screen shot from Fractal Dimension of the Box Data grid and the corresponding graph for the fractal line. The fractal dimension is the absolute value of the slope, in the middle bottom portion of the image.

If you choose to do the manual box counting method, instructions for that method are also included on the *Microbes Count!* web site.

Key Definitions:

Self-similar: This is the property such that a section of a shape appears like the original figure, regardless of scale – enlarging or reducing magnification.

Fractal: An object or quantity with a fragmented geometric shape that can be subdivided in parts, each of which (at least approximately) displays self-similarity. A natural fractal (also, random, stochastic) is a figure with self-similarity at all spatial scales and with fractional dimension: the object need not exhibit exactly the same structure at all scales, but the same "type" of structures must appear on all scales (Figure 4).

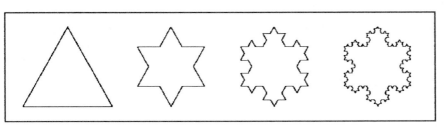

Figure 7. Barnsley's Fern and a Mandelbrot set

A deterministic (also, mathematical, non-random) fractal is a figure with self-similarity at all spatial scales (Figure 5).

Figure 8. A non-random fractal known as the Koch island.

John R. Jungck, Tia Johnson, and Joshua Tusin   The BioQUEST Curriculum Consortium

Euclidean dimension: A group of properties whose value (always an integer) is necessary and sufficient to determine uniquely each element of an object; this describes how points within an object are connected to each other. It specifies whether an object is a point, edge, surface, or solid. For example, a rectangle is two-dimensional, while a cube is three-dimensional.

Fractal dimension (Df): A group of properties whose value, usually non-integer, quantifies the degree to which the object fills the Euclidean space where the fractal set exists. It provides information about the length, area, or volume of an object. Mathematically, if a screen is laid over the fractal (for our purposes, in the second dimension) with screen mesh squares of size s, then the number of squares that cover part of the fractal figure is tallied as N boxes of size s: N(s). A plot of the quantity, on a log-log graph, versus scale gives a straight line, whose slope is the fractal dimension. $Df = \log(N(s))/\log(1/s)$.

The classical example for a fractal dimension is the length of a coastline measured with different length rulers. The shorter the ruler, the longer the length measured.

Diffusion-limited aggregation (DLA): This growth system begins with a single stationary seed particle, which is the nucleation site for a growth cluster. Moving particles are introduced and move about randomly. When a moving particle lands on a site adjacent to a stationary seed particle (an active site) it adheres (i.e., the moving particle aggregates and becomes a stationary seed particle). As particles continue to aggregate creating new cluster tips and edges, the probability to stick at the tips continually outweighs the probability to attach along the edges. This leads to branching. The probability for a particle to diffuse into the center of the growth cluster before encountering an active site becomes negligible as the cluster grows in size. Hence the outer tips grow more rapidly.

Acknowledgements:

Kevin Welch, a former undergraduate of Beloit College and Ph.D. student at Harvard University, Paul Trunfio, Department of Polymer Science, Boston University, Boston, MA, and Anton E. Weisstein, BioQUEST Curriculum Consortium.

Software Used in this Activity

Fractal Dimension
> Center for Polymer Studies, Boston University
>> Platform Compatibility: Macintosh and Windows

NIH Image
> Developed by Wayne Rasband at the U.S. National Institutes of Health and available for download at: http://rsb.info.nih.gov/nih-image/
>> Platform Compatibility: Macintosh only

Scion Image

Developed by Scion Corporation and available for download at:

http://www.scioncorp.com

Platform Compatibility: Windows only

Additional software listed in the "Analysis Procedures" document available on the *Microbes Count!* web site.

Additional Resources

Available on the *Microbes Count!* web site at http://bioquest.org/microbescount

Software

Fractal Dimension

Text

A PDF copy of this activity, formatted for printing

"Analysis Procedures" and "Box Counting Procedure"

Related *Microbes Count!* Activities

Chapter 1: Modeling More Mold

Chapter 7: Valuing Variegated Variation: Using Natural Experiments to Understand Viral Infection

Unseen Life on Earth Telecourse

Coordinates with Video II: The Unity of Living Systems

Relevant Textbook Keywords

Aesthetics, Bacterial morphology, Diffusion-limited aggregation (DLA), Fractal dimension

Related Web Sites (accessed on 4/19/03)

Center for Polymer Studies, Patterns in Nature
http://polymer.bu.edu/ogaf/

Eric Weisstein's World of Mathematics, information on fractals
http://mathworld.wolfram.com/Fractal.html

Information and images of fractal bacterial growth from the Bacterial Cybernetics Group, Tel Aviv University
http://star.tau.ac.il/~inon/baccyber0.html

Jacques Soddell and Robert Seviour, *Using Box Counting Techniques for Measuring Shape of Colonies of Filamentous Microorganisms*
http://www.csu.edu.au/ci/vol02/j_soddel/j_soddel.html

Microbes Count! Website
http://bioquest.org/microbescount

N. C. Kenkel and D. J. Walker, Quantitative Plant Ecology Laboratory, *Fractals in the Biological Science*
http://www.umanitoba.ca/faculties/science/botany/labs/ecology/fractals/fractal.html

Unseen Life on Earth: A Telecourse
http://www.microbeworld.org/htm/mam/is_telecourse.htm

References

Anonymous, (accessed 2003) *Fractal Geometry*
http://rsb.info.nih.gov/nih-image/download.html, go to Documents.

Ben-Jacob, E., I. Cohen, and D. Gutnick (1998). Cooperative organization of bacterial colonies: from genotype to morphotype. *Annual Review of Microbiology* 52:779-806.

Ben-Jacob, E. and H. Levine (1998). The artistry of microorganisms. *Scientific American* 279 (4):82-87 (October).

Ben-Jacob, E., O. Schochet, A. Tenenbaum, I. Cohen, A. Czirok, and T. Vicsek (1994). Generic modeling of cooperative growth patterns in bacterial colonies. *Nature* 368:46-49.

Center for Polymer Studies, Boston University (2001). Fractal Dimension. In *The BioQUEST Library Volume VI*. Jungck, J. R. and V. G. Vaughan, Editors. Academic Press: San Diego, CA.

Fujikawa, H. and M. Matsushita (1989). Fractal growth of Bacillus subtilis on agar plates. *J Phys Soc Japan* 58:3875-3878.

Fujikawa, H. and M. Matsushita (1991). Bacterial fractal growth in the concentration field of nutrient. *J Phys Soc Japan* 60:88-94.

Hartvigsen, G. (2000). The analysis of leaf shape using fractal geometry. *The American Biology Teacher* 62 (9):664-669 (November/December).

Trunfio, P. A. (self-published draft 11/12/93). *Pattern Formation by Bacterial Colonies*. Boston University, Department of Polymer Science: Boston, MA.

Bibliography

Bourke, P. (1993). Fractal Dimension Calculator, version 1.5. Auckland, New Zealand.

Jones, C. L., G. T. Lonergan, and D. E. Mainwaring (1993). A rapid method for the fractal analysis of fungal colony growth using image processing. *Binary* 5:171-180.

Kistler, R. A. (1995). Image acquisition, processing and analysis in the biology laboratory. *American Biology Teacher* 57 (3):151-157 (March).

Makse, H. A., S. Havlin, and H. E. Stanley (1995). Modeling urban growth patterns. *Nature* 377:608-612.

Matsushita, M. (1997) Formation of colony patterns by a bacterial cell population. J. A. Shapiro and M. Dworkin, eds. *Bacteria as Multicellular Organisms*. Oxford University Press:New York, NY.

Soddell, J., and R. Seviour (2000). *Using Box Counting Techniques for Measuring Shape of Colonies of Filamentous Microorganisms.* http://www.csu.edu.au/ci/vol02/j_soddel/j_soddel.html

Figure and Table References

Figure 1.   Bacterial Cybernetics Group, Ben-Jacob, et al. http://star.tau.ac.il/~inon/pictures/pictures.html

Figure 2.   Bacterial Cybernetics Group, Ben-Jacob, et al. http://star.tau.ac.il/~inon/baccyber0.html

Figure 3.   Bacterial Cybernetics Group, Ben-Jacob, et al. http://star.tau.ac.il/~inon/nature1/node7.html#SECTION00070000000000000000

Figure 4.   Courtesy of Tia Johnston

Figure 5.   Courtesy of Tia Johnston

Figure 6.   Courtesy of Tia Johnston

Figure 7.   Eric Weisstein's World of Mathematics http://mathworld.wolfram.com/Fractal.html

Figure 8.   Eric Weisstein's World of Mathematics http://mathworld.wolfram.com/Fractal.html

Code for generating high-resolution PostScript versions (and other formats) of the fractals in Figures 7 and 8 may be found in the Mathematica notebooks located at the following URLs:

http://mathworld.wolfram.com/notebooks/Fractals/KochSnowflake.nb

http://mathworld.wolfram.com/notebooks/Fractals/BarnsleyFern.nb

http://mathworld.wolfram.com/notebooks/Fractals/MandelbrotSet.nb

Chapter 3

Activities for Video III: Metabolism

It is surprising to learn that complex metabolic processes can be explored by looking at the products of populations of microbes. Here you can choose to look at the alcohol production in a wine vat, the pH of a yogurt culture, or the loss of oxygen in the Biosphere 2. The final activity pushes you to consider the complexities of microbial growth itself. You can explore microbial metabolism by controlling variables in the wet lab or by computer modeling and simulation.

In this unit, you can:

- model the fermentation process in early and modern wines,
- run simulations of a model of the Biosphere 2 to explore the unexpected impact of microbial metabolism on closed system dynamics,
- analyze serial dilution data from a living yogurt culture, and
- examine the dynamics of growth for populations of virtual bacteria with differing growth rates and carrying capacities.

Modeling Wine Fermentation

Humans have been producing wines for thousands of years. How did wine making get started? How has it changed? The Wine Mini-Model simulation enables us to explore the basic fermentation process as well as model enhancements such as the higher alcohol tolerance of cultivated yeasts used in modern wine making.

Biosphere 2: Unexpected Interactions

The Biosphere 2 project of the early 1990s failed dramatically because the planners overlooked the metabolic activities of naturally occurring soil microorganisms. The simulation program SimBio 2 allows us to gain insights into the complexity of this closed ecosystem. Here we can manipulate and monitor variables within a virtual Biosphere 2. We will explore oxygen availability by taking a closer look at the relationship between organic materials and living microbial populations in the soil. Alternative models can be constructed and simulated.

The Living World of Yogurt

Yogurt offers a non-pathogenic system in which to study microbial growth and metabolism. How do you estimate how many bacteria there are in a container of commercial yogurt? What are the effects on bacterial growth in fermenting yogurt with changes in pH, temperature, or substrate availability? You can even do the lab safely at home. Data from experimental runs are available for analysis.

Modeling Microbial Growth

Is bacterial growth always exponential? Do bacteria with the fastest rate of growth always have the largest populations? Biota models offer extended opportunities to observe population growth over time. What are the factors that affect growth? Explore continuous, chaotic, and cyclic growth models.

Modeling Wine Fermentation

Ethel D. Stanley, Howard T. Odum, Elisabeth C. Odum, and Virginia G. Vaughan

Video III: Metabolism

In the conversion of sugars into metabolic energy, yeasts and many bacteria produce ethanol instead of detoxifying it like we do. Sugars are broken down and used for energy, forming ethanol as the waste product, which is excreted by the cell. Grapes, with both low pH (2.8-3.8) and high sugar concentration (160 g/L to 240 g/L depending on variety and age), provide an ideal environment for the fermentative yeasts that are among the microbes found on the skins of grapes. Once the grapes are crushed, fermentation is probable, although a number of environmental variables can alter the process.

Figure 1. Wild populations of yeasts and bacteria thrive on the grapes as they are growing. When the fruit over-ripens, breaks in the skin allow these microbes entry and fermentation follows.

- Make a list of four variables you suspect would impact the fermentation of grapes.

- Assisting grape fermentation for the manufacture of wine has been a human endeavor since 6000 BC in Mesopotamia. Describe two procedures found in a modern winery that help ensure the success of the fermentation process. Be sure to identify the variables being controlled.

During fermentation, yeasts metabolize grape sugar for energy, producing carbon dioxide and ethanol as by products. The ethanol is transported rapidly outside of the yeast cells as a waste product. The yeast population grows quickly, grape sugar decreases, and ethanol accumulates to a level that begins to stress the yeast cells. In a closed batch process, the accumulation of ethanol eventually inhibits further microbial action and fermentation stops even though there may be sugar remaining.

- Before drawing blood samples, a phlebotomist swabs the insertion site on the patient's arm with an alcohol pad. Why?

We will further explore fermentation on the computer by:
- creating a simple model of wine fermentation,
- running a simulation with this model,
- carefully considering the results of this fermentation model, and
- investigating further "what if" questions by modifying the model.

Construct the Wine model

We will construct a simple model of wine fermentation using a modeling application called *Extend*. In an *Extend* model, the elements of the system are represented using picture icons and the flow of energy or products between the elements are represented using connecting lines. To create a model in *Extend* you add the appropriate icons to a worksheet and then connect them to create a system that is a simplified model of the real life system that you are studying. The mathematical equations that actually model the behavior of the wine fermentation system have been worked out in advance and are contained within the "blocks" that the picture icons represent.

The instructions below describe how to create the wine fermentation model. For more detailed instructions on creating, opening, running, saving, and modifying models see "Getting Started with *Extend*" on the *Microbes Count!* web site. The completed wine model is also available in the file called Wine.mox.

Extend LT

1. We are using a version of *Extend* called *Extend LT*. Start the *Extend LT* application by double-clicking on the application icon. The installer for the *Extend LT* application is in the Extend folder on the *Microbes Count!* web site.

2. Use File on the menu bar to Open the wine fermentation worksheet called WineWork.mox (located in the WineFermentation folder on the web site.)

3. The Wine worksheet opens up with an icon for a Plotter, which will be used to graph the results of your simulation.

4. To construct your model of wine fermentation, you will need to add icons for Grapes (grape sugar), Yeast, Ethanol, and Inhibition. To add the icons to your model, click and hold on the Library menu in the top menu bar. In the list of library files, move your mouse to WineLib.lix.

5. You will see a list of the icons available for this model. To put grapes into your model, click on the Grapes item. The Grapes icon will appear on your screen. You can click and drag the icon to arrange it on the worksheet. The cursor will become a hand when you drag an icon.

6. Add the Ethanol, Yeast, and Inhibition icons to the screen in the same way. You should arrange the icons so that your model looks similar to the model in Figure 2.

Figure 2. The completed wine fermentation model.

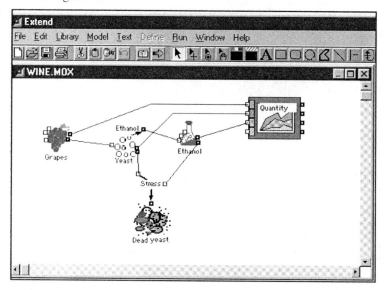

7. Each icon has connectors to attach it to other icons in your model (see the icons in Figure 3.) The small open box on the left of the picture icon is for the flow of energy or products into the icon, such as Grapes into the Yeast. The dark box on the right of an icon represents the flow of products or energy out of the icon, such as Ethanol from the Yeast. The Plotter has four boxes along the left side; it can keep track of four sets of changes in different units and draw four lines on the graph.

8. Now that you have added all of the icons to your worksheet you need to make the connections between them. With the mouse button held down, draw a line to connect the flow from the Grapes to the Yeast. As illustrated in Figure 3, the line goes from the right connector by the Grapes icon to the left connector of the Yeast icon.

   When you have successfully drawn a line between two connectors, the line becomes bold. To test the connection, drag the icon you just connected to. If the connection line follows, it is attached. If the line is not connected, delete it and try again. To delete a line or an icon, click on it to select it and then press the delete key.

9. Connect the Grapes, Yeast, Ethanol, and Inhibition (stress) icons to model the fermentation process. To make the program graph the changes in the amount of Grapes (or grape sugar), Yeast, and Ethanol you will also need to connect each to an output box left of the Plotter.

When you are finished your model should look like the model in Figure 2. Be sure you have connected the correct boxes; otherwise your model may not run properly. The completed model is also available in the file called Wine.mox.

Figure 3. Drawing a connection from the output box of the Grapes icon to the input box of the Yeasts icon. When the mouse is held over a connector box, it becomes a "pen" for drawing.

Run the simulation

Choose Run on the menu bar and highlight Run Simulation. A graph similar to the graph in Figure 4 appears. Grape sugar is scaled on the left vertical axis; yeast density and alcohol are scaled on the right vertical axis, referred to as Y2. Please note that at peak fermentation, there are between 1×10^6 and 25×10^6 yeast cells per milliliter in this model. The density value serves as an indicator for relative population growth.

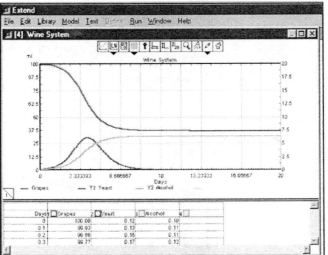

Figure 4. The results from running the wine fermentation simulation.

Ethel D. Stanley, Howard T. Odum, Elisabeth C. Odum, & Virginia G. Vaughan   The BioQUEST Curriculum Consortium

- Enter Day 4 and Day 5 values for grape sugar, yeast density, and alcohol in the table below.

  Note: You can scroll through the table below the graph or read quantities from the graph itself. To read quantities from a graph, move the mouse across the graph to the point you want to read. Then look at the top row of the table below the graph. For example, if you move the mouse across the graph to the Day 4, you can look at the top row of the table to see the values. Repeat for Day 5.

	Grape Sugar (ml)	Yeast Density	Alcohol (ml)
Day 4			
Day 5			
Comments			

- Are the values higher or lower on Day 5? Briefly explain why in the comments section.
- Does any grape sugar remain on Day 15? If so, how could you change in the model to use more of the grape sugar up?

Modify the model

One important distinction between wild yeasts and cultivated wine yeasts is their level of alcohol tolerance. Wild yeasts can tolerate about 4% alcohol. Then, as the alcohol concentration increases, fermentation slows down and stops. Tolerance to alcohol is higher in cultivated wine yeasts–up to 14%.

To simulate the improved alcohol tolerance of cultivated yeasts, you can reduce the stress of ethanol on the yeast population. To do this, double click on the Inhibition icon (see Figure 5.) Highlight the stress factor and enter 0.03 in the dialog box.

Figure 5. Modifying the wine model for reduced inhibition.

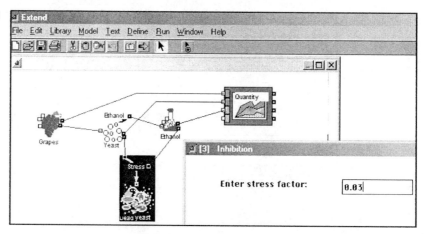

- How will the yeast population respond to this change?

- What will happen to the grape sugar and alcohol levels in the wine?

Run the revised model and compare the results to your predictions.

- Were your predictions accurate?

Explore more with the model called WineYeasts

In this model, two vats of grapes will be fermented under the same conditions at the same time, but with different yeasts (Figure 6.) The first vat, with untreated grapes, will simulate the use of a mix of wild yeasts for fermentation. The second vat, where sulfur dioxide was added early in the wine production to eliminate the wild yeasts, will simulate the use of *Sarccharomyces ellipsoideus*, a cultivated wine yeast with higher alcohol tolerance.

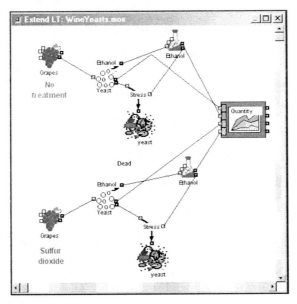

Figure 6. The WineYeasts model represents two vats with treated versus untreated grapes.

- Which yeasts will survive longer?

Now run the simulation and examine the graph and table to determine when the yeast population begins to decline. Note the total ethanol production.

- How could you use a similar model to convince winery owners to use the commercial yeasts your company makes?

Ethel D. Stanley, Howard T. Odum, Elisabeth C. Odum, & Virginia G. Vaughan   The BioQUEST Curriculum Consortium

Optional activities:

- Problem 1. Unlike the use of sourdough starters, "re-pitching" yeasts (saving yeasts from one batch to be re-used in another) is discouraged.

  Why is the first encouraged and the latter thought to be risky? You may want to look at the "Sourdough Symbiosis" activity in Chapter 2. (Hint: Explain the environmental factors that are less controlled in the latter.)

- Problem 2. In the model, you observed the relationships between grapes, yeasts and alcohol and discovered that the accumulation of a by-product (ethanol) inhibited further conversion of a substrate (grape sugar). The ethanol acts as a toxic waste product which eventually kills the yeasts. However, there are many metabolic pathways in which inhibition of a substrate conversion occurs without killing the organism involved.

  Use your text or other reliable resource to identify:

  1. A by-product found in another metabolic pathway which inhibits the conversion of a substrate without killing the organism involved.

  2. A toxic waste product and the organism that produces it.

Software Used in this Activity

Wine Fermentation

Howard T. Odum (University of Florida) and Elisabeth C. Odum (Santa Fe Community College)

Platform Compatibility: Macintosh and Windows

Additional Resources

Available on the *Microbes Count!* web site at http://bioquest.org/microbescount

Software

Wine Fermentation

Text

A PDF copy of this activity, formatted for printing

"Getting Started with *Extend*"

Related *Microbes Count!* Activities

Chapter 3: Modeling Wine Fermentation

Chapter 7: Microbiology of Stratified Waters

Chapter 8: Exploring Microbial Fermentation with Korean Kimchee

Chapter 10: Making Sense of Complex Life Cycles: Simulating Toxic Pfiesteria

Unseen Life on Earth Telecourse

Coordinates with Video III: Metabolism

Relevant Textbook Keywords

Ethanol, Fermentation, Metabolic Pathway, Substrate

Related Web Sites

American Home Brewing Supply - YEAST
http://www.redkart.com/cgi-bin/brew/ahbs/showstory.pl?s=newsletter&a=brew_palette_2

Microbes Count! Website
http://bioquest.org/microbescount

Office of Energy Efficiency and Renewable Energy
http://www.ott.doe.gov/biofuels/fermentation_background.html

Unseen Life on Earth: A Telecourse
http://www.microbeworld.org/htm/mam/is_telecourse.htm

Yeast
http://www.brewersworld.co.nz/Information/gervin.pdf

References

Beyers, R. J. and H. T. Odum (1993). *Ecological Microcosms.* Springer-Verlag: New York.

Evaluation of Yeast Viability and Concentration during Fermentation Using Flow Cytometry
http://www.bdbiosciences.com/immunocytometry_systems/application_notes/pdf/23-6289-01.pdf

Fermentations: Problems, Solutions and Prevention
http://www.uark.edu/depts/ifse/grapeprog/ articles/geis95wg.pdf

Henick-Kling, Thomas. Microbiology of Winemaking
http://www.nysaes.cornell.edu/fst/faculty/acree/fs430/lectures/thk05fermentation.html

Odum, H. T. and E. C. Odum (2000). *Modeling for All Scales, an Introduction to Simulation.* Academic Press, San Diego, CA.

Odum, H.T. and E.C. Odum (2001). *Wine Fermentation.*

PDB Newsletter: Alcohol Dehydrogenase
http://www.rcsb.org/pdb/newsletter/2001q1/mom.html

Wine Timeline
http://www.history-of-wine.com/html/timeline.html

Figure and Table References

Figure 1.   Courtesy Ethel D. Stanley

Figure 2.   Screen shot from *Wine Fermentation*

Ethel D. Stanley, Howard T. Odum, Elisabeth C. Odum, & Virginia G. Vaughan   The BioQUEST Curriculum Consortium

Biosphere 2: Unexpected Interactions

Ethel D. Stanley

Video III: Metabolism

Biosphere 2 is an enormous, glass-enclosed structure built in the mountains near Tucson, Arizona. The dome was designed as part of an experiment to create a self-sustaining environment capable of supporting scaled-down ecosystems including a forest, desert, and ocean. For two years (1991-1993), eight people lived within the dome and ran experiments tracking the interactions among land and water ecosystems and the atmosphere in great detail to help us better understand the ecology of the earth, our *Biosphere 1*. However, the rapid and unanticipated growth of microbes in the Biosphere 2 soil contributed to a steady decrease in oxygen, the loss of several species, and an early termination of the three year experimental project.

Figure 1. Inside Biosphere 2

The Biosphere 2 was designed as a simple balanced system in which oxygen, organic matter and carbon dioxide are produced, used, and recycled. The computer model, *SimBio2*, that you will be using in this activity was constructed as a "virtual" Biosphere 2. It uses the same basic processes considered in Biosphere 2 including production, consumption, and the recycling of materials

Figure 2. The Arizona desert is rich in organics.

You can use *SimBio2* to explore what went wrong. Start by opening the file "SB2Systm.mox", located in the Biosphere 2 folder on the *Microbes Count!* web site. You also have the option of assembling the model yourself by selecting icons from a library of Biosphere components (SimBioLibrary.LIX) and then connecting

Figure 3. The *SimBio2* model of Biosphere 2.

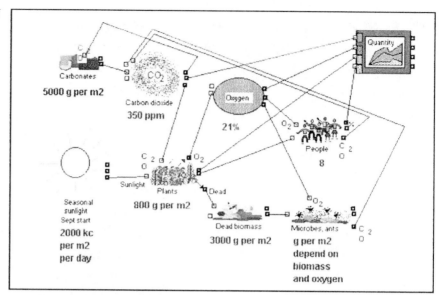

them as shown in Figure 3. The "Modeling Wine Fermentation" activity in Chapter 3 offers a detailed description of the process of creating a model. Additional information is available in the "Getting Started with *Extend*" document on the *Microbes Count!* web site.

In the *SimBio2* model, the elements of the system are represented using picture icons and the flows between the elements are represented using connecting lines. The icons in *SimBio2* (see Table 1) are used to represent:

- inflow of energy from the sun,

- photosynthesis by plants producing oxygen and biomass,

- buildup of dead biomass which is then consumed,

- consumption of plant biomass and oxygen by people, and

- recycling of carbon dioxide from all consumers both for use by plants and in storage as carbonates in seawater, carbonate-rich soil, and concrete.

In constructing the Biosphere 2, scientists underestimated the impacts of microbial growth. Microbes use oxygen in the process of consuming soil biomass for their food. When the soil is rich in organic biomass, the oxygen used in its decomposition can exceed the oxygen being produced by plants. In the consumption process, organic biomass and oxygen combine to make carbon dioxide and water.

Run the default model to explore how the amount of oxygen in the enclosed space of Biosphere 2 was reduced to a level that would not support humans and other species. (See Figure 4 for simulation results.) A graph traces the amounts of Plants and Carbon dioxide (scaled on the left vertical axis) as well as % Oxygen and % Life support (scaled on the right vertical axis referred to as Y2) over the time period selected.

Block	Description	Default Value
Seasonal sunlight	**Seasonal sunlight** This icon represents the average seasonal sunlight intensity, starting in September.	2000 kilocalories per square meter per day (kc/m^2)
Carbon dioxide	**Carbon dioxide** This icon represents the amount of gaseous carbon dioxide (CO_2) in Biosphere 2.	350 parts per million (ppm) by volume
Oxygen	**Oxygen** This icon represents the percent of oxgen (O_2) in Biosphere 2.	21 percent by volume of gases in the air
Sunlight Plants Dead	**Plants** This icon represents the plant biomass (grasses, shrubs, trees and other plant producers) in Biosphere 2.	800 grams per square meter (g/m^2)
Carbonates	**Carbonates** This icon represents carbonates and bicarbonates stored in limestone rocks, soil, water and concrete.	5000 grams per square meter (g/m^2)
People	**People** This icon represents the people's life support, which is a combination of food and oxygen needed. The plotter records the percent of necessary diet being supplied.	8 people.
Dead biomass	**Dead biomass** This icon represents the quantity of dead plant litter (leaves, twigs, branches), detritus, animal feces, and dead animals in Biosphere 2.	3000 grams per square meter (g/m^2)
Microbes, ants	**Microbes, ants, etc.** These organisms are decomposers, using the dead biomass for food.	Depends on oxygen and dead biomass. Grams per square meter (g/m^2)

Table 1. Descriptions and default values of the icons in the *SimBio2* model.

Figure 4. Results of running the *SimBio2* model, with added labels.

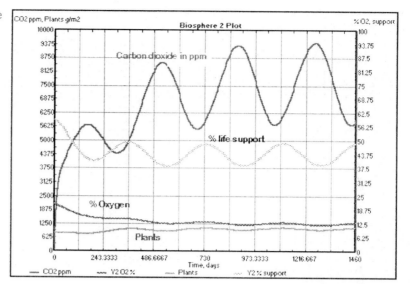

Revising the model

SimBio2 provides "hands on" opportunities for you to learn more about the Biosphere 2 experiment through the design and testing of experimental models. The use of models is a routine and necessary part of scientific research. Researchers are able to explore what they think they already know as well as what they don't know. You can explore the *SimBio2* model by running simulations, looking at the experimental results, and revising the model. You can very the amount of time that the simulation runs or change starting values of the components.

You can use *SimBio2* to investigate what is likely to happen if the variables are changed. For example, what if a different soil was used in the Biosphere 2?

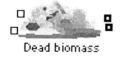

Dead biomass

To simulate the use of forest soil with less organics, you need to reduce the organic matter by half. Double-click on the dead biomass icon and change the starting value from 3000 g/m2 to 1500 g/m2.

- Monitor the O_2 and CO_2 levels. What is different from the default simulation results?

- Describe how the level of starting soil fertility (dead biomass) impacts the level of oxygen.

- Modify the starting value for dead biomass and run the simulation. Do these new results support your hypothesis above? Explain.

Further Investigations

- What alternative strategies can you offer to keep the oxygen level from dropping? What if you increased the amount of plants? Decreased the number of people in the dome? Increased the starting % of oxygen in the dome?

- What variable(s) have you chosen to explore?

- How will you change the model?

- What do you think will happen to the oxygen level?

Report your results.

Extended problem:

Scientists recently completed a three-year experiment in which two areas of an aspen forest in Wisconsin were maintained with the atmospheric CO_2 at ambient levels in one and ambient-plus-200 ppm in the other. Not only did the trees respond favorably to the increase in CO_2, but also the microbes in the soils underlying the trees. Phillips et al. reported "microbial respiration was 29% greater beneath plants growing under elevated CO_2."

- Offer an explanation for these experimental results.

- What kinds of information would you gather to support your explanation?

- How might modeling be useful in explaining the results as well as setting up new experiments?

Software Used in this Activity

SimBio2—Simulating Biosphere 2
Howard T. Odum (University of Florida) and Elisabeth C. Odum (Santa Fe Community College)

Platform Compatibility: Macintosh and Windows

Additional Resources

Available on the *Microbes Count!* web site at http://bioquest.org/microbescount

Software
SimBio2—Simulating Biosphere 2

Text
A PDF copy of this activity, formatted for printing

"Getting Started with *Extend*"

Related *Microbes Count!* Activities

Chapter 3: Modeling Wine Fermentation

Chapter 7: Microbiology of Stratified Waters

Chapter 10: Making Sense of Complex Life Cycles: Simulating Toxic Pfiesteria

Unseen Life on Earth Telecourse

Coordinates with Video III: Metabolism

Relevant Textbook Keywords

Biomass, Biosphere, Ecosystem, Organics

Related Web Sites (accessed on 4/5/03)

Biochemists Solve Mystery of Lost Oxygen
http://www.columbia.edu/cu/record/record1931.23.html

Columbia University Biosphere 2
http://www.bio2.edu/

For further simulations and other questions to explore, go to:
http://bioquest.org/simbio2.html

Historical Overview of the Biosphere 2 Project
http://www.biospheres.com/allenasa.html

Microbes Count! Website
http://bioquest.org/microbescount

Overview of the Biosphere 2 complex - its main features and a short history
http://helios.bto.ed.ac.uk/bto/microbes/biosph.htm

Unseen Life on Earth: A Telecourse
http://www.microbeworld.org/htm/mam/is_telecourse.htm

References

Biosphere's Response to Higher CO_2
http://www.greeningearthsociety.org/Articles/2002/biosphere.htm

Odum, E.C. and H.T. Odum (2001). *SimBio2–Simulating Biosphere 2*. In *The BioQUEST Library Volume VI*, Jungck, J. R. and V. G. Vaughan, Editors. Academic Press: San Diego, CA.

Phillips, R. L., D. R. Zak, W. E. Holmes, and D. C. White (2002). Microbial community composition and function beneath temperate trees exposed to elevated atmospheric carbon dioxide. *Oecologia*, 131: 236-244.

Figure and Table References

Figure 1.   Photo courtesy of Ethel D. Stanley

Figure 2.   Photo courtesy of Ethel D. Stanley

Figure 3.   Screen shot from *SimBio2*

Figure 4.   Screen shot from *SimBio2*

The Living World of Yogurt

Marion Field Fass

Fermented milk products have been made for centuries by people who wished to retain milk's food value despite its tendency to rapidly spoil. Many different species of lactic-acid producing bacteria turn camel's milk, goat's milk or ewe's milk into local products. In 1915 Elie Metchnikoff isolated two species of bacteria that are responsible for true cow's milk yogurt, *Lactobacilli bulgaris* and *Streptococcus lactis*. Metchnikoff believed that the longevity of people from Bulgaria and the Caucasus mountains was due to their consumption of yogurt. Today, Metchnikoff's belief lives on in those who purchase "probiotics" such as lactobacilli for their positive effects on health.

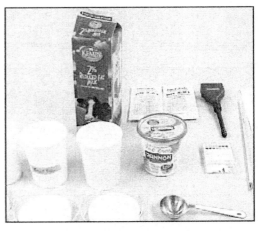

Figure 1. Supplies for the cultivation of *Lactobacilli bulgaris* and *Streptococcus lactis* in the laboratory.

The process by which milk becomes yogurt provides a model for microbiological investigation. This is one of the activities in *Microbes Count!* that uses non-pathogenic organisms to study metabolic processes and that allows safe, open-ended investigations of bacterial metabolism in the laboratory and at home. "Exploring Microbial Fermentation with Korean Kimchee" in Chapter 8 is another.

This laboratory investigation is derived from William Coleman's "Koch's Postulates and Yogurt" (1995). Students use Koch's postulates to guide the identification of the "agent" that causes milk to develop the disease we call yogurt. Complete instructions can be found in Coleman's article, included on the *Microbes Count!* web site.

You can proceed with the wet laboratory or use the data provided on the *Microbes Count!* web site to estimate how many bacteria are found in a container of yogurt and to explore how the pH of milk changes after the addition of active yogurt cultures.

Overview of the procedure

While we know that bacteria are very small, it is hard to imagine how many living bacteria there are in a container of yogurt. Many commercially available yogurts advertise "Contains live cultures." Using techniques of serial dilution and bacterial culture, we will test the validity of this claim and estimate how many bacteria the yogurt actually contains.

Our method of counting bacteria by serial dilution assumes that if we spread a very thin or dilute layer of bacteria on a petri plate filled with agar containing the appropriate nutrients, the distinct colonies that are visible after incubation each result from one individual cell. There are too many bacteria to count in undiluted yogurt, so we will have to estimate by taking a small sample (0.1 ml) and seeing how many bacteria there are in that sample. In order to count bacteria in a sample, we will smear the yogurt evenly on a plate of media selected to grow lactobacilli and count how many colonies appear.

If there are too many bacteria in our 0.1 ml sample to count, we can increase the accuracy of our estimate by using less yogurt. We will do this by diluting the yogurt further, to a 1:100 dilution. We will then plate 0.1 ml of this mixture on the plate of media and incubate to see what grows.

There may still be too many bacteria in the 1:100 dilution to count accurately. If so, we will dilute the yogurt even farther, to a 1:10,000 dilution. We will then plate 0.1 ml of this mixture to see how many colonies grow. We assume that each colony that appears on a plate is the result of one bacterial cell in the original batch. The count on the final plate is one tenth of a 10,000 dilution.

Finally, we will make our own yogurt and observe the effects of different temperatures on the pH of the milk as it becomes yogurt.

Determining the number of bacteria in a container of yogurt by serial dilution and plating

In the laboratory: Serial dilutions

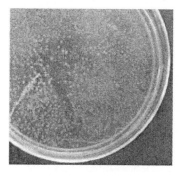

Figure 2. A plate from a 1/100 dilution.

1. Using pH paper or an automated sensor, determine the pH (acidity) of the undiluted yogurt.

2. Place 0.1 ml of the undiluted yogurt on an agar plate. Using a sterile spreader, spread the yogurt evenly across the agar.

3. Add 0.1 ml yogurt to a 9.9 ml sterile blank. Mix. This is a 1/100 dilution.

4. Place 0.1 ml of the diluted yogurt (1/100) on an agar plate. Spread, using a sterile spreader.

5. Add 0.1 ml of the mix from Step 3 to a 9.9 ml sterile blank. Mix. This is a 1/10,000 dilution.

6. Place 0.1 ml of the diluted (1/10,000) yogurt on an agar plate. Spread.

7. Label the plates with the name of your group and the dilution and incubate aerobically or anaerobically at 43-46 degrees Celsius for 24-36 hours.

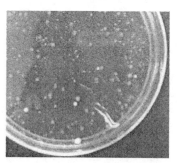

Figure 3. A plate from a 1/10,000 dilution.

In the classroom or at home: Estimating

1. Count colonies on each of your plates and write the number on the board or share by email with classmates. Some plates may have too many colonies to count. Alternatively, use the data in Table 1 for your calculations. (The data in Table 1, and all of the other tables in this activity, are also available on the *Microbes Count!* web site.)

2. Calculate the mean number of colonies on undiluted plates, 1:100 plates and 1:10,000 plates. Calculate standard deviation as well. If there are too many plates that have too many colonies to count, you may decide that this is not an accurate dilution for estimation.

Table 1. Serial Dilution Data

Sample Set of Student Lab Data						
SERIAL DILUTION DATA						
	Pure Culture, Undiluted		Diluted Culture 1:100		Diluted Culture 1:10,000	
	Colonies per cm^2	Colonies per Plate	Colonies per cm^2	Colonies per Plate	Colonies per cm^2	Colonies per Plate
Group 1	117	7020	56	3360	45	2700
	130	7800	66	3960	33	1980
	134	8040	86	5160	25	1500
	54	3240	82	4920	15	900
Group 2	Uncountable	Uncountable	Lawn	Lawn	3.5	213
	Uncountable	Uncountable	Lawn	Lawn	14.2	852
Group 3	Uncountable	Uncountable	Lawn	Lawn	4	240
	Uncountable	Uncountable	Lawn	Lawn	14	840
	Uncountable	Uncountable	Lawn	Lawn	7	420
Group 4	Uncountable	Uncountable	Lawn	Lawn	31.6	1896
	Uncountable	Uncountable	Lawn	Lawn	16.4	984
	Uncountable	Uncountable	Lawn	Lawn	24	1440
Group 5	Uncountable	Uncountable	Lawn	Lawn	18	1080
	Uncountable	Uncountable	Lawn	Lawn	27	1620
Group 6	189	11340	26	1560	6	360
	207	12420	40	2400	9	540
Group 7	Uncountable	Uncountable	Lawn	Lawn	11	660
	Uncountable	Uncountable	Lawn	Lawn	12	720
	Uncountable	Uncountable	Lawn	Lawn	14	840

3. Using the 1:10,000 dilution, calculate how many bacteria there are in one 8 ounce container of yogurt. Since we plated only 0.1 ml of diluted yogurt, this 1:10,000 dilution really represents how many colonies there are in 0.1 ml of the diluted mix. To estimate the number in 1 ml of your dilution, multiply the average number of colonies by 10.

4. If you can calculate how many colonies are in 1ml of 1:10,000 dilution, you can then figure out how many colonies are in 1 ml of undiluted yogurt. Calculate this.

5. If you can estimate how many bacteria there are in 1 ml of yogurt, you can calculate how many there are in a whole container.

    How many milliliters are there in an 8 ounce container? What is your estimate for a container of yogurt? Does the estimate you can make from your plates differ from the estimate that you make from the averaged class data? Why? Which do you think is more accurate?

Making yogurt and examining environmental effects on bacterial metabolism

In the laboratory:

1. Use a commercially available yogurt maker or a collection of clean storage jars to make your own yogurt. The standard method is to scald the milk (heat to just below boiling) and then cool to 40 degrees Celsius. Add 2 tablespoons of unflavored commercial yogurt and mix well. Pour this warm milk mixture into 5 clean containers. Incubate in a yogurt maker, refrigerator, or waterbath.

2. Using a pH meter, test the changes in pH of milk infected with commercial yogurt as it changes. Measure every hour for at least 8 hours or longer if possible. How does temperature affect changes in pH? Does the pH continue to fall?

3. If you cannot measure pH for a long period, you can use the data in Table 2 below. What are the optimum temperatures for growth of these bacteria?

Figure 4. Measuring pH

Table 2. Yogurt Lab Data (no nutrients)

Yogurt Lab Data (0 - 12 hr)								
Waterbath ~45°C			Yogurt Maker ~32°C			Refrigerator ~ 4°C		
Hour	pH	Sample Temp	Hour	pH	Sample Temp	Hour	pH	Sample Temp
0	5.493	39	0	5.461	39	0	5.844	39
1	5.391	37.56	1	5.393	36.7	1	6.05	13.68
2	4.946	38.4	2	5.097	34.57	2	6.107	13.11
3	4.761	41.31	3	4.9	34.48	3	6.075	11.06
4	4.464	40	4	4.716	37	4	6.183	7.7
5	4.339	40.83	5	4.597	36.99	5	6.114	6.24
6	4.191	40.98	6	4.452	37.59	6	6.177	5.33
7	4.07	39.34	7	4.334	36.72	7	6.215	5.29
8	3.999	40	8	4.256	37.76	8	6.237	5.89
9	3.964	40.99	9	4.154	37	9	6.203	5.37
10	3.902	41.07	10	4.082	37.44	10	5.623	5.39
11	3.904	38.96	11	4.049	37.1	11	6.216	4.96
12	3.843	40.11	12	4.005	37.05	12	6.196	5.39

4. What hypotheses can you make about the factors that limit the growth of yogurt producing bacteria in milk? The data in Table 3 show what happens when non-fat dry milk or sucrose are added to yogurt cultures in milk. What do these data tell you about the reproduction of bacteria?

    (Note: In the commercial preparation of yogurt, sugar and flavorings are added after fermentation. Why do you think this is?)

5. Use a spreadsheet program such as Microsoft Excel® to graph the changes in pH in the yogurt samples. How do they change?

Table 3. Yogurt Lab Data (with added nutrients)

Yogurt Lab Data (Added Nutrients; Data 0 - 14 hr)											
Sugar - Waterbath			Milk - Waterbath			Sugar - Yogurt Maker			Milk - Yogurt Maker		
Hour	pH	Temp	Hour	pH	Temp	Hour	pH	Temp	Hour	pH	Temp
0	5.40	34.13	0	5.58	31.54	0	5.42	32.76	0	5.61	32.98
1	5.05	42.45	1	5.05	43.72	1	5.28	36.68	1	5.54	33.80
2	4.65	43.30	2	4.75	43.90	2	5.09	36.34	2	5.36	33.98
3	4.41	43.28	3	4.54	44.62	3	4.86	36.64	3	5.03	35.99
4	4.21	43.76	4	4.35	45.07	4	4.63	36.67	4	4.83	34.84
5	4.13	43.72	5	4.21	44.48	5	4.47	35.47	5	4.72	37.01
6	3.97	43.73	6	4.08	44.42	6	4.28	36.84	6	4.47	37.47
7	3.97	41.07	7	4.01	42.32	7	4.20	36.58	7	4.49	37.51
8	3.88	42.15	8	4.03	42.84	8	4.13	37.38	8	4.48	36.84
9	3.75	43.15	9	3.93	43.90	9	4.06	35.96	9	4.33	37.14
10	3.75	43.01	10	3.87	43.91	10	3.96	36.34	10	4.22	37.32
11	3.78	42.81	11	3.87	44.63	11	3.96	35.52	11	4.21	37.27
12	3.70	42.55	12	3.80	43.88	12	3.85	35.41	12	4.17	37.54
13	3.66	42.01	13	3.77	43.90	13	3.83	35.84	13	4.17	38.00
14	3.63	41.90	14	3.76	43.42	14	3.81	35.92	14	4.11	37.69

Questions:

• What factors affect the growth of bacteria in yogurt?

• Will the bacteria continue to keep metabolizing milk?

• How low will the pH go?

- Would the addition of other nutrients result in different pH levels?

- What would yogurt taste like if the acidity fell to pH 3?

- What factors limit bacterial growth in yogurt?

Additional Resources

Available on the *Microbes Count!* web site at http://bioquest.org/microbescount

Text

A PDF copy of this activity, formatted for printing

Table 1, Table 2, and Table 3 in PDF format for printing

Original data for Table 1, Table 2, and Table 3, plus additional data

William Coleman's "Koch's Postulates and Yogurt" (1995), in PDF format. (Included with permission of *BioScene: Journal of College Biology Teaching* http://acube.indstate.edu/ .)

Related *Microbes Count!* Activities

Chapter 1: How Do You Know a Microbe When You See One?

Chapter 2: Sourdough Symbiosis

Chapter 3: Modeling Wine Fermentation

Chapter 8: Exploring Microbial Fermentation with Korean Kimchee

Unseen Life on Earth Telecourse

Coordinates with Video III: Metabolism

Relevant Textbook Keywords

Aerobe, Anaerobe, Koch's Postulates, *Lactobacilli*

Related Web Sites (accessed 2/22/03)

Microbes Count! Website
    http://bioquest.org/microbescount

Unseen Life on Earth: A Telecourse
    http://www.microbeworld.org/htm/mam/is_telecourse.htm

Yogurt and Health
    http://www.fda.gov/bbs/topics/CONSUMER/CON00150.html

References

Coleman, W. H. (1995). Koch's postulates and yogurt. *BioScene, 21*, 3–6.
(A PDF version of this article is available on the *Microbes Count!* web site
and on the web at http://acube.org/volume_21/v21-2p3-6.pdf .)

Figure and Table References

Figure 1.   Courtesy Tia Johnson and Joshua Tusin

Figure 2.   Courtesy Tia Johnson and Joshua Tusin

Figure 3.   Courtesy Tia Johnson and Joshua Tusin

Figure 4.   Courtesy Tia Johnson and Joshua Tusin

Table 1.   Serial Dilution Data provided by Tia Johnson and Kari Roettger

Table 2.   No Nutrient Data provided by Tia Johnson and Kari Roettger

Table 3.   Added Nutrient Data provided by Tia Johnson and Kari Roettger

Modeling Microbial Growth

Ethel D. Stanley

Video III: Metabolism

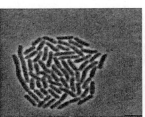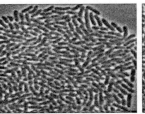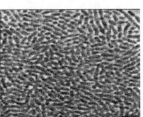

Figure 1. Bacteria reproduce exponentially when conditions are favorable. This population of hundreds of organisms all arose from a single bacterium.

One of the most remarkable features of microbes is their potential for rapid population growth. Sometimes this growth surprises us- for example, the carbonate-rich desert soil and favorable environmental conditions in the Biosphere 2 dome supported the reproduction of microbes. As escalating populations of bacteria used oxygen within the dome, members of several larger species – notably the birds - suffered irreparable physiological damage and died. Biosphere scientists monitoring the unexpected decrease in oxygen were puzzled. They had overlooked the reproductive potential of microbes.

Do microbial populations always reproduce exponentially? The answer is no. Most populations grow more slowly because the conditions for their growth are not ideal. You should be familiar with several ways that you can interrupt the exponential growth of microbes–e.g. refrigerating your milk, washing your hands, or applying antiperspirant.

An enormous amount of data has been collected on microbial growth. "Most studies in food microbiology are concerned with the rapid growth of populations." McMeekin (1997) emphasizes the role of quantitative microbial ecology and modeling.

> *"Predictive microbiology involves knowledge of microbial growth responses to environmental factors summarized as equations or mathematical models. The raw data and models may be stored in a database from which the information can be retrieved and used to interpret the effect of processing and distribution practices on microbial proliferation. Coupled with information on environmental history during processing and storage, predictive microbiology provides precision in making decisions on the microbiologic safety and quality of foods."*

In this exercise, you will explore models of growth in order to understand the problems associated with microbial sampling in environmental and applied microbiology. Suppose you are a fermentation technician at an industrial processed food laboratory. Your job involves the maintenance of cultures of bacteria under optimal conditions. One major task is the harvesting of microbial by-products from these cultures. You must monitor bacteria and decide when to harvest.

You are going to need more information. What does a population growth curve look like? What factors do you need to take into account? Sometimes the accumulation of by-products results in the death of the culture, so you might want to begin harvesting when the growth rate drops. (See the "Modeling Wine Fernentation" activity in Chapter 3.)

To answer these and other questions, you will need to monitor the populations. This is difficult to do in real life because of complicating factors such as limited nutrient supply, immunological responses, predators, or even restrictive physical conditions like temperature and pH changes. Since it is hard to measure the growth in situ for microbial populations, we will take a closer look using different growth models developed in a software application called *Biota*.

Explore the Logistic Continuous Growth model

Double click the *Biota* icon to begin. An installer for *Biota* is located in the Modeling Growth folder on the *Microbes Count!* web site. The startup window will open with several choices. Click on Open Document to bring up a list of all simulations, problem files and folders in the current directory. Choose the Logistic Continuous Growth Model.

Figure 2. *Biota* screen shots of the Map and Growth Parameters.

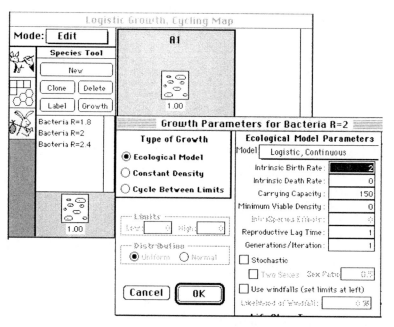

Note: The logistic model, first described by Verhulst in 1838, assumes the rate of population growth depends on the population density when all other conditions are constant. For example, in a flask containing nutrient media that is recently inoculated with *E. coli*, the growth rate of the population is close to or at maximum. As this population increases towards the maximum number of cells that can be supported by the available media, the growth rate slows. If the number of cells exceed this carrying capacity, the growth rate can fall below zero. (This model is continuous; i.e. nutrients are replaced in the flask so that the available resources are held at a constant level.)

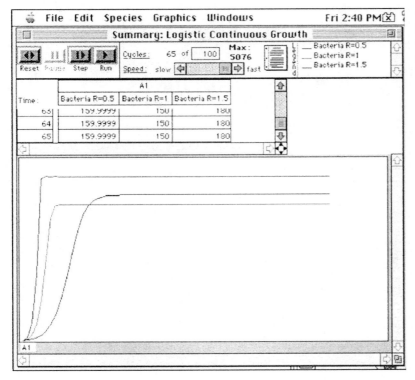

Figure 3. Simulation results for three populations of bacteria with different growth rates. All three populations begin with a lag phase that slows and then levels off. Note that the population curves on the graph have different shapes.

Run this simulation to compare "ideal" logistic growth by three bacteria. The first species reproduces half as fast as the second species and a third as fast as the last species. Figure 3 shows the results of the simulation.

- What differences do you see based only on the reproductive rates?

- The carrying capacity (the maximum number of individuals in a population that can be supported by the environment) clearly differs for each species. Which population has the fewest individuals at carrying capacity?

- If you were asked to recommend two of these bacteria for harvesting a byproduct whose production is strongly correlated with population size, which would it be? Why?

Note: This "continuous" growth model is used for populations that reproduce asexually. Not surprisingly, sexually reproducing populations require a more sophisticated growth model such as Lotka-Volterra that depends on discrete generations and generates much slower increases in population size. Rabbits, rotifers, and rhododendrons require a different model than bacteria.

Explore the Logistic Growth Chaotic Model

Environmental factors must also be considered when you are trying to determine the size of a population. Although you are growing cultures under controlled conditions in a lab, keep in mind that most environments for living organisms are not ideal. A more realistic model for looking at population growth includes stochastic (chaotic) events like a drought or a chemical spill.

Figure 4. Simulation results for two populations of bacteria with similar growth rates.

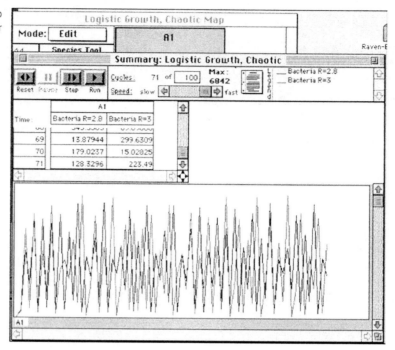

Run this simulation for two independent bacteria with similar reproductive rates and note the results.

Run the simulation four more times. Look carefully at the data tables as well as the graphs for each run.

- Do bacteria with the faster reproductive rate always have the largest population size?

Explore the Logistic Growth Cyclic model

Microbial environments change over time and populations may be quite cyclic as they increase and decrease in response to environmental factors such as insufficient food or pressure from an increase in predator population.

Figure 5. Screenshot of the Map for the Logistic Growth model featuring three bacteria with different growth rates.

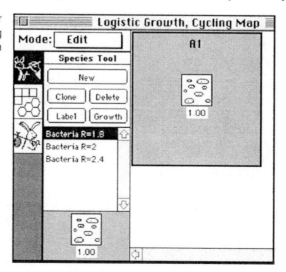

- Using your text, a reliable internet source, or scientific journals, provide at least two specific examples of cycling microbial populations and a brief explanation of the environmental factor limiting each.

Before you run this next simulation featuring three bacteria with differing reproductive rates, predict which of these populations you expect to be the largest in the first cycle of growth.

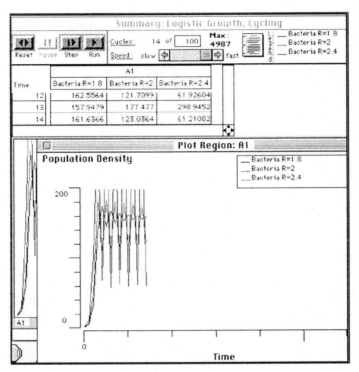

Figure 6. Sample simulation results for Logistic Growth model.

- Are the populations as uniform as those produced by continuous growth? (You can double click on the graph itself to open a new window that can be increased in size by dragging a corner.)

- How does the population graph you generated compare to your chaotic logistic growth results?

- Explain how you could use the simulation results above to argue for daily rather than weekly water quality testing for an area of public swimming in urban lakes.

Optional Activity

- Using selected data (Veilleux 1976) from a graduate student thesis investigating the growth of *Didinium* (table on the right), the following graph was produced for a population grown in a 5 cubic centimeter culture.

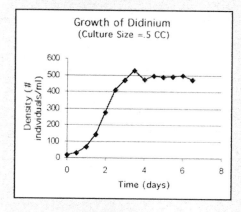

Growth of Didinium
(Culture Size =.5 CC)

Time (days)	Density (# individuals/ml)
0.0	16
0.5	30
1.0	66
1.5	142
2.0	275
2.5	410
3.0	469
3.5	526
4.0	472
4.5	497
5.0	489
5.5	492
6.0	497
6.5	473

- How do the results above compare with your Biota simulation results?

The data for a .375 cubic centimeter culture follows. Graph the result and compare these two populations.

- Do you have new insights about population growth and its relationship to available resources?

Time (days)	Density (# individuals/ml)
0	17
0.5	25
1.0	43
1.5	74
2.0	133
2.5	230
3.0	356
3.5	393
4.0	401
4.5	363
5.0	403
5.5	375
6.0	385
6.5	377

Software Used in this Activity

Biota

Jim Danbury, Ben Jones, John Kruper, Eric Nelson, William Sterner, Jeff Schank, Jim Lichtenstein, Joyce Weil, and William Wimsatt (University of Chicago)

Platform Compatibility: Macintosh only

Additional Resources

Available on the *Microbes Count!* web site at http://bioquest.org/microbescount

Software

Biota

Text

A PDF copy of this activity, formatted for printing

Related *Microbes Count!* Activities

Chapter 1: Population Explosion: Modeling Phage Growth

Chapter 3: Biosphere 2: Unexpected Interactions

Chapter 8: Modeling Microbial Predator-Prey Relationships

Chapter 9: Mold Fights Back: A Challenge For Sanitation

Relevant Textbook Keywords

Asexual, Logistic growth (log phase)

Related Web Sites

Current Research Involving Microbial Growth:
  http://es.epa.gov/ncer/final/centers/hsrc/89/western/mccarty11_98.html

Data set: Prey species: *Paramecium aurelia*; Predator species: *Didinium nasutum*
  http://www.inapg.inra.fr/ens_rech/bio/Ecologie/fichiers/PRSLBDataSets.txt

Logistic Growth Model
  http://www.math.duke.edu/education/ccp/materials/diffeq/logistic/logi1.html

References

Danbury, J., B. Jones, J. Kruper, E. Nelson, W. Sterner, J. Schank, J. Lichtenstein, J. Weil, W. Wimsatt 2001. Biota. In *The BioQUEST Library Volume VI*, Jungck, J. R. and V. G. Vaughan, Editors. Academic Press: San Diego, CA.

T. A. McMeekin, J. Brown, K. Krist, D. Miles, K. Neumeyer, D.S. Nichols, J. Olley, K. Presser, D. A. Ratkowsky, T. Ross, M. Salter, and S. Soontranon (1997). Quantitative Microbiology: A Basis for Food Safety. *Emerging Infectious Diseases*. Vol 3(4).
http://www.cdc.gov/ncidod/eid/vol3no4/mcmeekin.htm

Veilleux, B. (1976) The analysis of a predatory interaction between *Didinium* and *Paramecium*. Thesis. University of Alberta.

Figure and Table References

Figure 1.   Courtesy Ethel D. Stanley

Figure 2.   Screen shot from *Biota*

Figure 3.   Screen shot from *Biota*

Figure 4.   Screen shot from *Biota*

Figure 5.   Screen shot from *Biota*

Figure 6.   Screen shot from *Biota*

Chapter 4

Activities for Video IV: Reading the Code of Life

In the fifty years since the structure of DNA was first published, public awareness of DNA has soared. We live at a time when the genome for the emerging SARS virus could be accessed over the internet within weeks of its isolation from patients. Now that genomics and proteomics have entered the public arena, it is clear that educated citizens of the 21st century will need to understand the essential roles of DNA.

In this unit we can:

- recreate classic experiments in the discovery of DNA,
- use sequence information to develop a protocol for treating antibiotic resistant TB,
- examine viral DNA sequence data as forensic evidence, and
- explore a data set of HIV nucleic acid sequences utilizing bioinformatics tools and resources.

Searching for the Hereditary Molecule

How was the hereditary molecule isolated and identified? What decisions do we need to make as we recreate the lab work of researchers like Oswald Avery and his colleagues? Using the simulation Searching for the Hereditary Molecule, we can manipulate virtual tubes of bacteria in a virtual lab to discover the transforming principle ourselves.

TB and Antibiotic Resistance

How does the development of antibiotic resistant strains of TB influence modern healthcare practices? We will consider genetic, environmental, epidemiological, and social perspectives of this renewed threat to public health. The TB simulation allows us to experiment with several TB strains and the antibiotics used to treat TB as we explore antibiotic resistance.

Molecular Forensics

Can we establish the origin of an infection by looking at DNA? In the 1990's, suspicion that HIV in several individuals could be linked to a local dentist was investigated. We will take a closer look at sequence data from this dentist and other HIV positive individuals including patients who believed they were exposed to HIV during dental procedures.

Exploring HIV Evolution: An Opportunity for Research

What can differences in the DNA of a rapidly evolving virus tell us? A data set of nucleic acid sequences for strains of HIV collected from patients over a three year period provides an opportunity to use bioinformatics to study the natural history of disease. Here we can develop and test our own hypotheses about virulence and life expectancy.

Searching for the Hereditary Molecule

Marion Field Fass and Donald Buckley

Video IV: Reading the Code of Life

This activity allows you to simulate the classic experiments that demonstrated how genetic information is transferred in the process of transformation. First you must imagine that you were working in the years before 1952 when Watson and Crick described the structure of DNA.

Background

A long series of related and unrelated experiments provided the background information that led Watson and Crick to propose a molecular mechanism for the transfer of genetic information. Critical to this background was Fred Griffith's 1920 experiment on the bacteria that caused a pneumonia epidemic in Britain. Griffith, an epidemiologist, had found that the pneumococci that infected patients were of two different phenotypes—smooth and rough—and that only the smooth ones were harmful to patients they infected.

Figure 1. Oswald Avery, 1937

In his now classic experiment, Griffith injected mice with the rough bacteria, and found that the mice survived. He then injected mice with the smooth bacteria, and the mice got sick and died.

Griffiths then injected mice with living rough bacteria, and heat killed smooth bacteria. He found that mice injected with this mixture died. When Griffith cultured the bacteria from the dead mice, he found that he could culture both living smooth bacteria and living rough bacteria. Somehow the rough bacteria had picked up traits from the dead smooth bacteria and expressed these new traits. Griffith called this process "transformation."

Two decades of active scientific advances passed before Oswald Avery, a bacteriologist at Rockefeller Institute, refined Griffith's experiment to try to specify what had caused the transformation that enabled the bacteria to kill the mice they infected. He knew, based on some experiments by Colin MacLeod, that temperatures up to 80° C destroyed the outer cell wall of the bacteria, but left the DNA, RNA and protein intact. Higher temperatures denatured these cellular components.

Avery and his colleagues had a range of tools that weren't available to Griffith in the 1920s. Scientists had improved techniques for bacterial culture, and had isolated enzymes that could digest protein, RNA and DNA. By the 1940s, Avery was able to study the changes that took place in bacteria in Griffith's experiment without using the mice as "incubators." Avery and his colleagues were able to study the bacteria directly on petri plates.

The Simulation

Avery, with Colin MacLeod and Maclyn MacCarthy, undertook a series of experiments to determine what substance–RNA, DNA or protein–was responsible for the transformation of the bacteria. Using the computer application *The Search for the Hereditary Molecule*, you can repeat the experiments that took them years to perfect. You will now have the same materials that the scientists did: smooth and rough bacteria, media to grow them in, enzymes to digest DNA, RNA and protein, and lots of petri plates on which to grow bacteria.

Open *The Search for the Hereditary Molecule* application. Click anywhere on the opening page to move to the lab bench (Figure 2).

Figure 2. A screen shot of the lab bench in *The Search for the Hereditary Molecule* application.

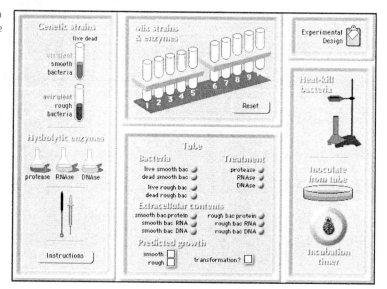

To set up your first experiment, drag the pipette to fill the test tubes with bacteria and enzymes. If you want to use heat killed bacteria, drag the test tube over to the bunsen burner, and then back again. Make sure to make predictions about whether you will be able to grow rough or smooth bacteria from the mixture you make. To see your result, use the loop to inoculate the petri plate on the right, and set the timer. (Click on the Instructions button on the lower left for additional information on using *The Search for the Hereditary Molecule* application.)

Another way to set up an experiment is to use the pull down menu at each tube number to directly add bacteria and enzymes to each tube. To see your result, use the loop to inoculate the petri plate on the right, and set the timer.

The check marks will indicate if your prediction was right or wrong.

Rarely can a scientist support a conclusion based on the results of a single experiment. Usually there are multiple experiments repeated multiple times. You can click on the clipboard on the upper right of the desktop to get to a worksheet for setting up multiple experiments.

Questions

1. Prior to Avery's experiments, scientists were confused about the role of DNA in cells. Watson and Crick were not yet on the scene. Write a letter to the editor of a scientific journal reporting the experiments you have completed and what they tell you about the role of DNA.

2. While Avery's experiments seem simple to us now, the scientists faced many technical challenges. Do research to determine some of the challenges faced by this team of scientists. What were they?

3. Today we understand that DNA is the hereditary molecule responsible for the transfer of new traits in transformation. There are many things we don't know about transformation, however. Review recent articles in *Science*, *Science News* or *Scientific American* to identify some of the unanswered questions. What are they? Distinguish between questions of theory and those of techniques. How do the methods that scientists use to approach these questions differ?

Software Used in this Activity

Search for the Hereditary Molecule

Donald Buckley (Quinnipiac University) and William Coleman (University of Hartford)

Platform Compatibility: Macintosh only

Additional Resources

Available on the *Microbes Count!* web site at http://bioquest.org/microbescount

Software

Search for the Hereditary Molecule

Text

A PDF copy of this activity, formatted for printing

Related *Microbes Count!* Activities

Chapter 5: Conjugation and Genetic Mapping

Chapter 6: One Cell Three Genomes

Unseen Life on Earth Telecourse

Coordinates with Video IV: Reading the Code of Life

Relevant Textbook Keywords

Bacterial culture, DNA, Protein, RNA, Transformation

Related Web Sites (accessed on 2/21/03)

DNA, RNA, and Protein
http://press2.nci.nih.gov/sciencebehind/genetesting/genetesting07.htm

Microbes Count! Website
http://bioquest.org/microbescount

Transformation discovery
http://profiles.nlm.nih.gov/CC/A/A/O/J/

The Oswald T. Avery Collection
http://profiles.nlm.nih.gov/CC/

Unseen Life on Earth: A Telecourse
http://www.microbeworld.org/htm/mam/is_telecourse.htm

References

Buckley, D. and W. Coleman (2001). Search for the Hereditary Molecule: Avery Transforms the Search. In *The BioQUEST Library Volume VI*. Jungck, J. R. and V. G. Vaughan, editors. Academic Press: San Diego, CA.

Hagen, J. B. (1996). Oswald Avery and the Search for the Transforming Factor. In Hagen, J.B., D. Allchin, and F. Singer *Doing Biology*. Harper Collins College Publishers: New York. pp 60-70.

Figure and Table References

Figure 1.   Permission of the Rockefeller University Archives
http://www.rockefeller.edu/archive.ctr/

Figure 2.   Screen shot from *Search for the Hereditary Molecule: Avery Transforms the Search*

TB and Antibiotic Resistance

Marion Field Fass

Video IV: Reading the Code of Life

The ability of bacteria like *Streptococcus* or *Mycobacterium tuberculosis* to develop resistance to antibiotics provides a tremendous challenge to microbiologists and health care professionals. Antibiotics use a range of strategies to kill bacteria—some destroy the cell walls of bacteria, while others interfere with the metabolic activities of the cell.

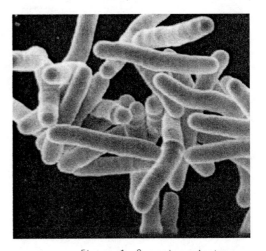

Figure 1: Scanning electron micrograph of Mycobaterium tuberculosis.

There are 5 major targets of antibiotic action (Talaro and Talaro, 2002):

1. Inhibition of cell wall synthesis.

    Examples: penicillin, bacitracin, cephalosporin

2. Disruption of cell membrane function.

    Example: polymyxin

3. Inhibition of protein synthesis.

    Examples: tetracycline, erythromycin, streptomycin, chloramphenicol

4. Inhibition of nucleic acid synthesis.

    Examples: rifamycin (transcription), quinolones (DNA replication)

5. Action as anti-metabolites.

    Examples: sulfonilamide, trimethoprim

But how do bacteria acquire resistance to antibiotics? Clearly, in the game of life, a bacterium that can resist the killing action of an antibiotic has a selective advantage. Random mutations may give rise to genetic traits that confer resistance, or genetic traits can be transferred between bacteria (conjugation) through the sharing of plasmids. A variety of genetically encoded traits may give bacteria the ability to resist the effects of antibiotics.

In this activity you have the opportunity to analyze different strains of *Mycobacterium tuberculosis* to see which of several antibiotics provides the best treatment. You will then be able to perform several chemical tests to investigate why a strain of the bacteria has developed resistance to a particular antibiotic. These are important questions. If you know that a strain of TB is susceptible, or resistant, to a specific antibiotic, you can minimize the exposure of the patient to an ineffective drug. On a population level, you can limit the development of further antibiotic resistance by carefully managing your drug use. Understanding how a drug works to kill the bacteria is important to understanding the mutations that lead them to fail.

The situation is this: You are the microbiologist in charge of the lab at a large urban hospital. Recently six patients have been admitted to your hospital with active tuberculosis. They face long and difficult recovery periods, in which they will need to take anti-tuberculosis drugs for 6 months to 12 years. Because the *Mycobacterium tuberculosis* that causes TB are such slow growing bacteria, it can take at least six months for the drugs to kill all the bacteria in a person's body. People have difficulty adhering to anti-TB drugs. Since adherence is key in controlling antibiotic resistance, many public health districts have instituted a program called DOTS–Directly Observed Therapy–in which a health worker visits the patient daily to assure that he/she is taking the drugs as prescribed. Because this is an expensive and time consuming process, your job is to learn as much as possible about the six strains of *M. tuberculosis* that are infecting the patients.

The *TB Lab: Antibiotic Resistant Bacteria* software application provides a simulated laboratory for your investigations (Figure 2). In this lab you can grow bacteria with antibiotics, run chemical tests to explore how antibiotics affect the cellular processes of the *M. tuberculosis* bacteria, and compare parts of the DNA sequences of the different strains of bacteria. Information on using *TB Lab* is available on the *Microbes Count!* web site.

Figure 2. The *TB Lab* workbench.

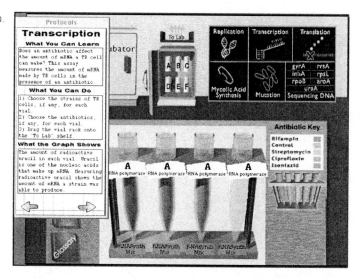

1. You have been asked to analyze which drugs effectively treat each strain of tuberculosis in your patient population. Can you identify which strains are resistant to the usual selection of drugs used to treat TB?

2. As a scientist, you are interested in finding out the modes of action of each of the antibiotics in your cabinet. How do these work to kill the *Mycobacterium tuberculosis*? Would you want to prescribe more than one antibiotic to a patient? Why?

3. The doctors and nurses on the floor are concerned that several of the patients might have trouble following a daily regimen of taking their antibiotics. What do you think would be the consequences of a patient taking their pills for only four weeks, until he/she felt better, rather than completing the six

month course of treatment? What would be the consequences for a patient who took the medicine sporadically-perhaps only once or twice a week when he/she remembered?

4. The DNA of *Mycobacterium tuberculosis* holds the instructions for the synthesis of cell walls, transcription of DNA, and all other important life functions. By analyzing DNA sequences, scientists can identify differences in bacterial metabolism that may confer resistance to antibiotics. Use the DNA sequences provided in the *TB Lab* and the results of your tests of the antibiotic susceptibility or resistance of the TB strains to suggest the relationship between mutations and drug activity for the strains of TB that infect the patients in your hospital.

5. The County Health Officer has also contacted you, as head of the lab, about the genetic sequences of the strains of TB that the patients have. The County Health Officer would like to use this information to identify chains of infection in the community. How do the genetics of the different strains of TB from the patients in your lab differ? Are there any patients who, based on the genetic analysis of their TB strains, you think could have been infected from the same source? Why? Describe the similarities and differences between the genetic sequences of the TB strains in your lab. What implications do these have for effective treatment of TB?

Software Used in this Activity

TB Lab–Antibiotic Resistant Bacteria
William Sandoval, Brian J. Reiser, Renee Judd, and Richard Leider (Northwestern University)

Platform Compatibility: Macintosh only

Additional Resources

Available on the *Microbes Count!* web site at http://bioquest.org/microbescount

Software
TB Lab—Antibiotic Resistant Bacteria

Text
A PDF copy of this activity, formatted for printing

Related *Microbes Count!* Activities

Chapter 6: Tracking the West Nile Virus

Chapter 9: Mold Fights Back: A Challenge for Sanitation

Chapter 11: Identifying Immune Responses

Unseen Life on Earth

Video IV: Reading the Code of Life

Relevant Textbook Keywords

Antibiotic resistance, Conjugation, Mutation, *Mycobacterium tuberculosis*, Plasmid, Strain, Tuberculosis

Related Web Sites (accessed on 2/21/03)

Alliance for the Prudent Use of Antibiotics
http://www.healthsci.tufts.edu/apua

Antibiotic Resistance
http://www.fda.gov/fdac/features/795_antibio.html
http://www.niaid.nih.gov/director/congress/1999/0225.htm

Centers for Disease Control, Division of TB Elimination
http://www.cdc.gov/nchstp/tb/default.htm

Making Directly Observed Therapy Work. Khalil Sabu Rashidi and Debra Bottinick, MPH.
www.umdnj.edu/ntbcweb/mdotw.html

Microbes Count! Website
http://bioquest.org/microbescount

Multidrug-Resistant Tuberculosis Fact Sheet. American Lung Association
www.lungusa.org/diseases/mdrtbfac.html

Tuberculosis
http://science-education.nih.gov/nihHTML/ose/snapshots/multimedia/ritn/Tuberculosis/tb1.html

United States Food and Drug Administration
http://www.fda.gov

Unseen Life on Earth: A Telecourse
http://www.microbeworld.org/htm/mam/is_telecourse.htm

References

Sandoval, W., B. J. Reiser, R. Judd, and R. Leider. (2001). BGuILE: TB Lab-Antibiotic Resistant Bacteria. In *The BioQUEST Library Volume VI*, Jungck, J. R. and V. G. Vaughan, Editors. Academic Press: San Diego, CA.

Talaro, K. P. and A. Talaro (2002). *Foundations in Microbiology.* McGrawHill: New York, NY. Page 351.

Figure and Table References

Figure 1.   Courtesy of Clifton E. Barry, III and Elizabeth Fischer (Laboratory of Immunogenetics, NIAID, National Institutes of Health)

Figure 2.   Screen shot from *TB Lab–Antibiotic Resistant Bacteria*

Molecular Forensics

Sam Donovan

Video IV: Reading the Code of Life

Introduction

Molecular data are routinely used to determine paternity and to link physical evidence at a crime scene to individuals. The specificity of DNA sequence data makes it possible to link tissue samples—even very small amounts of hair, blood, saliva or semen—to the individual who is the source of the material, with a high degree of confidence and accuracy. The idea of a DNA "fingerprint" involves comparing a set of "molecular markers" from a tissue sample to samples from known subjects. The identity, or lack of identity, between the known and unknown samples can then be used as evidence in a court of law.

In addition to using molecular data to find identity matches, DNA evidence has been used in courtrooms in other ways as well. In this exercise we will look at how the analysis of viral evolution can be used to make biological and legal arguments about the transmission of the virus between individuals. Instead of looking for an identity match you will use the similarity between sequences to group strains of the virus and make inferences about their historical relationships. We know that a population of HIV viruses within a person evolves quickly due to its rapid generation time and frequent mutations. We can use the pattern of these changes to look for evidence that people share the same virus source.

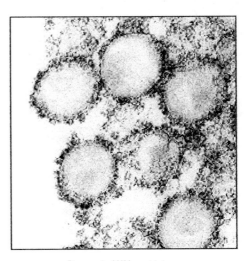

Figure 1. HIV particles

This activity is based on an actual case which broke new ground with respect to the use of molecular evidence in a courtroom. While the exercises outlined here use some of the actual data collected and analyzed as part of the court proceedings, they do not address the complexity nor the depth of the evidence. They do, however, provide you with an opportunity to practice some of the reasoning used to go from sequence data to biological inference, and to translate that biological understanding into evidence that could be used in a court of law. For more information about the history of this case see the Additional Resources section at the end of this activity.

Background

In the spring of 1990, Kimberly, a 22 year-old living in Fort Pierce, Florida, tested positive for HIV. This was surprising because she had no identifiable risk factors for contracting the virus. Epidemiological research focused on an invasive dental procedure performed by an HIV positive dentist several years earlier. Searching the dentist's records revealed a number of other HIV positive patients, several of whom also had no known risk factors for contracting the virus. The

Centers for Disease Control and Prevention (CDC) became involved and the case received a great deal of media attention due to the public's concern that HIV+ health care workers might be a threat to their patients. Multiple lawsuits were filed by patients claiming that they were infected by the dentist and seeking damages.

The following exercises provide an introduction to working with sequence data and drawing inferences from sequence alignment and distance trees. By comparing the genetic sequences of HIV virus genes isolated from blood samples from the dentist, his patients, and other HIV+ individuals in the community who did not have contact with the dentist (local controls), scientists worked to determine if there was a relationship between the dentist's and patients' viruses. As you try to establish whether there is evidence that the dentist was responsible for the HIV infection in his patients you will also explore:

- characteristics of nucleic acid sequence data;

- how a multiple sequence alignment can be used to visualize similarities and differences between sequences; and,

- how a distance graph (unrooted tree) represents the differences between sequences and can be used to develop hypotheses about their evolutionary relationships.

Part 1 - Comparing Raw Sequence Data

The data in Table 1 below includes 3 sequences from different patients (E, F and G), a dentist sequence, and 2 sequences from HIV+ individuals who lived in the area but had not had contact with the dentist (local controls). Each sequence is from the same part of the HIV genome (the GP120 gene, V3 region) — a variable region of a gene that makes a viral coat protein.

Study the data in Table 1 and discuss the questions below with your group members. Be prepared to share your findings and any questions that arise.

- What sorts of patterns do you see within/between these sequences?

- How are these sequences similar (different)?

- Are they all similar/different in the same ways?

- How do you think this information could be used to determine if the dentist was the source of the HIV in the patients?

Dentist	GAGGTAGTAATTAGATCTGCCAATTTCACAGACAATGCTAAAATCATAATAGTACAGCT GAATGCATCTGTAGAAATTAATTGTACAAGACCCAACAACTATACAAGAAAAGGTATA CGTATAGGACCAGGGAGAGCAGTTTATGCAGCAGAAAAAATAATAGGAGATATAAGA CGAGCACATTGTAACATTAGTAGAGAAAAATGGAATAATACTTTAAAACAGGTAGTTA CAAAATTAAGAGAACAATTTGTGAATAAAACAATAATCTTTACTCACCCCTCAGGAGGG GACCCAGAAAT
Patient E	GAGATAGTAATTAAATCTGCCAATTTCACAGACAATGCTAAAATCATAATAGTACAGCT GAATGCATCTGTAGAAATTAATTGTACAAGACCCAACAACAATACAAGAAAAGGTATA CATATAGGACCAGGGGAGGGCATTTTATGCAACAGGAGAAATAATAGGAGATATAAGAC AAGCACATTGTAACATTAGTGGAGAAAAATGGAATAATACTTTAAAACAGGTAGTTAC AAAATTAAGAGAACAATTTGGGAATAAAACAATAATCTTTAATCACTCCTCAGGAGGG GACCCAGAAAT
Patient F	GAAGTAGTAATTAGATCTGAAAAATTTCACGGACAATGTTAAAACCATAATAGAGCAGC TGAATGAATCTGTACAAATTAATTGTACAAGACCCAACAACAATACAAGAAAAAGTAT ACATATAGCACCGGGGGAGAGCATTTTATGCAACAGGAGAAATAATAAGAGATATAAGA CAAGCACATCGTAACCTTAGTAGCATAAAAATGGAATAACACTTTAAGACAGATAGCTAA AAAATTAAAAGAACAATTTGGAAATAAAACAATAATCTTTAATCAATCCTCAGGAGGG GACCCAGAAAT
Patient G	GAGGTAGTAATTAGATCTGCCAATTTCACAGACAATGCTAAAATCATAATAGTACAGCT GAATGCACCTGTAGAAATTAATTGTACAAGACCCAACAACAATACAAGAAAAGGTATA AGTATAGGACCAGGGGAGAGCATTTTATGCAACAGATAGAATAGTAGGAGATATAAGAA AAGCATATTGTAACATTAGTAGAGAAAAATGGAATAATACTTTAAAACTGGTAGTTAC AAAATTAAGAGAACAATTTGTGAATAAAACAATAATCTTTAATCACTCCTCAGGAGGG GACCCAGAAAT
Local Control 3	GAGGTAGTAATTAGATCTGAAAAATTTCACGGACAATACTAAAACCATAATAGTACAGCT AAATACATCTGTAACAATTAATTGTACAAGACCTGGCAACAATACAAGAAAAAGTATA ACTATGGGACCGGGGGAAAGTATTTTATGCAGGAGAAATAATAGGAGATATAAGACAAG CACATTGTAACCTTAGTAGAACAGCATGGAATGACACTTTAGAACAGATAGTTGGAAA ATTACAAGAACAATTTGGGAATAAAACAATAGTCTTTAATCACTCCTCAGGAGGGGACC CAGAAAT
Local Control 22	GAGGTAGTAATTAGATCTGACAATTTCTCGGACAATGCTAGAACCATAATAGTACAGCT GAACGAATCTGTAGTAATTAATTGTACAAGACCCAACAACAATACGAGCAGACGTATA AGTATAGGACCAGGGGAGAGCATTTACTGCAAGAGAAGGAATAATAGGAGACATAAGA CAAGCACATTGTAACATTAGTGGAGCAGAATGGGAAAGCACTTTAAAACGGATAGTTG AAAAATTAGGAGAACAATTTAAGAATAAAACAATAGTCTTTAATCACTCCTCAGGAGG GGACCCAGAAAT

Table 1: HIV nucleic acid sequences from 6 subjects.

Part 2 - Interpreting a Multiple Sequence Alignment

While it is possible to manually compare raw sequence data, it quickly gets unwieldy when you are working with long sequences or many different sequences. Luckily, computers are very efficient at following instructions and performing mathematical operations. In this section you will interpret the output from a program that has performed a multiple sequence alignment on the six sequences you looked at in Part 1. The ClustalW program is used to "align" sequences by finding the best ways to make their different nucleotide positions line up with one another and then color coding the positions (columns) to characterize the types of differences between sequences. Figure 2 shows part of a ClustalW multiple sequence alignment of the raw sequence data from Table 1.

Figure 2. Multiple sequence alignment.

```
Sequence alignment

Consensus key (see documentation for details)
* - single, fully conserved residue
  - no consensus


CLUSTAL W (1.81) multiple sequence alignment

Dentist     GAGGTAGTAATTAGATCTGCCAATTTCACAGACAATGCTAAAATCATAATAGTACAGCTG
Patient_F   GAAGTAGTAATTAGATCTGAAAATTTCACGGACAATGTTAAAACCATAATAGAGCAGCTG
Patient_G   GAGGTAGTAATTAGATCTGCCAATTTCACAGACAATGCTAAAATCATAATAGTACAGCTG
Patient_E   GAGATAGTAATTAAATCTGCCAATTTCACAGACAATGCTAAAATCATAATAGTACAGCTG
L_C_3       GAGGTAGTAATTAGATCTGAAAATTTCACGGACAATACTAAAACCATAATAGTACAGCTA
L_C_22      GAGGTAGTAATTAGATCTGACAATTTCTCGGACAATGCTAGAACCATAATAGTACAGCTG
            **  ********* ***** ****** * ****** ** ** ******** *****

Dentist     AATGCATCTGTAGAAATTAATTGTACAAGACCCAACAACTATACAAGAAAAGGTATACGT
Patient_F   AATGAATCTGTACAAATTAATTGTACAAGACCCAACAACAATACAAGAAAAAGTATACAT
Patient_G   AATGCACCTGTAGAAATTAATTGTACAAGACCCAACAACAATACAAGAAAAGGTATAAGT
Patient_E   AATGCATCTGTAGAAATTAATTGTACAAGACCCAACAACAATACAAGAAAAGGTATACAT
L_C_3       AATACATCTGTAACAATTAATTGTACAAGACCTGGCAACAATACAAGAAAAAGTATAACT
L_C_22      AACGAATCTGTAGTAATTAATTGTACAAGACCCAACAACAATACGAGCAGACGTATAAGT
            **   * *****  ***************** ****  **** ** * * *****  *

Dentist     ATAGGACCAGGGAGAGCAGTTTATGCAGCAGAAAAAAATAATAGGAGATATAAGACGAGCA
Patient_F   ATAGCACCGGGGGAGAGCATTTTATGCAACAGGAGAAATAATAAGAGATATAAGACAAGCA
Patient_G   ATAGGACCAGGGAGAGCATTTTATGCAACAGGATAGAATAGTAGGAGATATAAGAAAAGCA
Patient_E   ATAGGACCAGGGAGGGCATTTTATGCAACAGGAGAAATAATAGGAGATATAAGACAAGCA
L_C_3       ATGGGACCGGGGAAGTATTTTATGCA---GGAGAAATAATAGGAGATATAAGACAAGCA
L_C_22      ATAGGACCAGGGAGAGCATTTACTGCAAGAGAAGGAATAATAGGAGACATAAGACAAGCA
            ** * *** ****  * * **  ****   *     **** ** **** ******  ****
```

   • Does the information presented in the multiple sequence alignment support
     the patterns you saw when you looked at the raw sequence data?

   • Why do you think that the Local Control 3 (L_C_3) sequence had a "gap" (-)
     inserted in its sequence?

   • How do you think this information could be used to determine if the dentist
     was the source of the HIV in the patients?

Part 3 - Reading a Distance Tree

Another way to compare these sequences is to look at the genetic similarity
between each pair of sequences. In this section this pairwise similarity is reported
first as a table of pairwise % identities and then as a distance tree. The % identity
table was generated by comparing each sequence with every other sequence,
calculating the number of positions that are identical and then dividing that number
by the total number of positions. The distance tree represents genetic distances
between sequences as lengths between the tips of the branches. Thus, tracing the
lines from one sequence, or branch tip, to another correlates with the pairwise
distance reported in Figure 3.

Part of determining if the dentist is the source of the patients' HIV is seeing how
the sequences group together based on their similarity. Because HIV evolves
rapidly, it is very unlikely that the viral sequences in two subjects would be
identical even if one person had transmitted the virus to the other. However, using
the assumption that genetically similar sequences are more closely related to one

another and therefore share a more recent common ancestor, it is possible to infer past identity from similarity.

```
CLUSTAL W (1.81) Multiple Sequence Alignments


Sequence type explicitly set to DNA
Sequence format is Pearson
Sequence 1: Dentist      302 bp
Sequence 2: Patient_F    302 bp
Sequence 3: Patient_G    302 bp
Sequence 4: Patient_E    302 bp
Sequence 5: L_C_3        299 bp
Sequence 6: L_C_22       302 bp
Start of Pairwise alignments
Aligning...
Sequences (1:2) Aligned. Score:  87
Sequences (1:3) Aligned. Score:  95
Sequences (1:4) Aligned. Score:  95
Sequences (1:5) Aligned. Score:  86
Sequences (1:6) Aligned. Score:  86
Sequences (2:3) Aligned. Score:  87
Sequences (2:4) Aligned. Score:  89
Sequences (2:5) Aligned. Score:  88
Sequences (2:6) Aligned. Score:  84
Sequences (3:4) Aligned. Score:  94
Sequences (3:5) Aligned. Score:  86
Sequences (3:6) Aligned. Score:  87
Sequences (4:5) Aligned. Score:  87
Sequences (4:6) Aligned. Score:  87
Sequences (5:6) Aligned. Score:  86
```

Figure 3: Pairwise sequence similarities

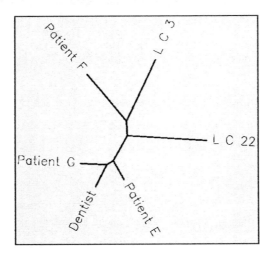

Figure 4: Distance tree.

- Does the information presented in the table of pairwise similarity scores (Figure 3) and in the distance tree (Figure 4) support the patterns you saw when you looked at the raw sequence data and the multiple sequence alignment?

- Why do you think some of the lines are longer than others?

- Do you think the places where the lines connect with one another is important? How can the internal branches be interpreted?

- How do you think this information could be used to determine if the dentist was the source of the HIV in the patients?

Assignment

- Write a brief summary of the argument you would make to a judge or jury regarding the claims from these three patients that they received their HIV infections from the dentist. Refer directly to the data available in this exercise and be explicit about how you are interpreting it. Be careful not to overstate your conclusions.

Critical thinking questions

- What does the distance tree diagram tell you about the directionality of HIV transmission? What does it tell you about the likelihood of direct transmission between dentist and patient(s)? Does your language in the argument above reflect this understanding?

- This exercise has been simplified dramatically by including only one sequence from each subject in the analysis. In fact, each of these individuals had a variable population of HIV in their bodies. How might information about the similarities and differences between HIV populations for each individual shape your argument?

- Explain the importance of having local controls in your analysis. How are these sequences used to inform the case you make in the courtroom?

Web Resources Used in this Activity

Biology Workbench (http://workbench.sdsc.edu)

The *Biology Workbench* was originally developed by the Computational Biology Group at the National Center for Supercomputing Applications at the University of Illinois at Urbana-Champaign. Ongoing development of version 3.2 is occurring at the San Diego Supercomputer Center, at the University of California, San Diego. The development was and is directed by Professor Shankar Subramaniam.

Platform Compatibility: Requires an internet connection and a current browser.

Additional Resources

Available on the *Microbes Count!* web site at http://bioquest.org/microbescount

Text

A PDF copy of this activity, formatted for printing

Teaching Notes

Dataset from lab, for use with the Teaching Notes

Related *Microbes Count!* Activities

Chapter 2: Searching for Amylase

Chapter 4: Exploring HIV Evolution: An Opportunity for Research

Chapter 6: Proteins: Historians of Life on Earth

Chapter 6: Tree of Life: Introduction to Microbial Phylogeny

Chapter 6: Tracking the West Nile Virus

Chapter 6: One Cell, Three Genomes

Chapter 7: Visualizing Microbial Proteins

Unseen Life on Earth Telecourse

Coordinates with Video IV: Reading the Code of Life

Relevant Textbook Keywords

Amino acids, Bioinformatics, Disease, Epidemiology, Evolution, HIV, Nucleic
acids, Phylogenetic relationships, Virus

Related Web Sites

Biology Workbench
http://workbench.sdsc.edu

Forensics and Genetics
http://ornl.gov/hgmis/elsi/forensics.html

Microbes Count! Website
http://bioquest.org/microbescount

Unseen Life on Earth: A Telecourse
http://www.microbeworld.org/htm/mam/is_telecourse.htm

References

Popular literature
Gentile, B. (July 1, 1991). Doctors with AIDS. *Newsweek* 48-56.

Scientific Literature
Ou, C. Y., C. A. Ciesislski, G. Myers, et al. (1992). Molecular epidemiology
of HIV transmission in a dental practice. *Science* 256:1165–1171.
(PMID: 1589796; UI: 92271245)

Bibliography

Barr, S. (1996). The 1990 Florida dental investigation: Is the case really closed?
Annals of Internal Medicine 124:250-254.
http://www.acponline.org/journals/annals/15jan96/flordent.htm

The Centers for Disease Control and Prevention (CDC) publish a weekly newsletter called the Morbidity and Mortality Weekly Report (MMWR). The following articles contain nice concise descriptions of the work involved in understanding the transmission of HIV in this case.

Possible Transmission of Human Immunodeficiency Virus to a Patient during an Invasive Dental Procedure. July 27, 1990, 39(29):489-493
http://www.cdc.gov/mmwr/preview/mmwrhtml/00001679.htm

Epidemiologic Notes and Reports Update: Transmission of HIV Infection during an Invasive Dental Procedure–Florida. January 18, 1991, 40(2):21-27, 33
http://www.cdc.gov/mmwr/preview/mmwrhtml/00001877.htm

Epidemiologic Notes and Reports Update: Transmission of HIV Infection During Invasive Dental Procedures–Florida. June 14, 1991, 40(23):377-381.
http://www.cdc.gov/mmwr/preview/mmwrhtml/00014428.htm

Investigations of Persons Treated by HIV-Infected Health-Care Workers–United States. May 07, 1993, 42(17):329-331,337
http://www.cdc.gov/mmwr/preview/mmwrhtml/00020479.htm

Figure and Table References

Figure 1.   www.ncbi.nlm.nih.gov

Figure 2.   Modified from Biology Workbench (http://workbench.sdsc.edu)

Figure 3.   Modified from Biology Workbench (http://workbench.sdsc.edu)

Figure 4.   Modified from Biology Workbench (http://workbench.sdsc.edu)

Table 1.   Sequences obtained from GenBank (http://www.ncbi.nlm.nih.gov/)

Exploring HIV Evolution: An Opportunity for Research

Sam Donovan and Anton E. Weisstein

Video IV: Reading the Code of Life

Human Immunodeficiency Virus (HIV), like other retroviruses, has a much higher mutation rate than is typically found in organisms that do not go through reverse transcription (the copying of RNA into DNA). Mansky and Temin (1995) estimated the rate of point mutations in HIV to be 3×10^{-5} errors per base per replication cycle. The impact of HIV as a pathogen is due in large part to this high mutation rate which, among other things, causes the surface proteins to change and avoid normal immune detection and suppression. A great deal has been learned about the evolution of HIV because it is relatively easy to sample the rapidly changing population of viruses within an infected individual and look at the patterns of molecular change over time.

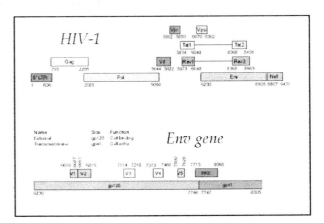

Figure 1: Schematics of the HIV-1 genome (top) and *env* gene (bottom). Note the overlap in the coding regions of genes. The variable regions (V1-V5) of *env* gene are loops that extend from the core of the protein.

In this activity you will study aspects of sequence evolution by working with a set of HIV sequence data from 15 different subjects (Markham, et al., 1998). You will first learn about the dataset, then study the possible sources of HIV for these subjects, and then design and pursue your own research project.

Basic HIV biology such as life history characteristics of the virus and its interactions with the human immune system are not discussed here but are important background for placing these exercises into a broader biological and clinical context. A few suggested resources for reviewing basic HIV biology are provided in the reference section.

An orientation to the HIV sequence data

The HIV genome is very small and relatively simple. It is made up of nine genes and about 9,500 nucleotides. In this lab, you will use sequences from the envelope gene, *env* (see Figure 1). The envelope gene codes for two membrane proteins (gp41 and gp120) that extend from the cell membrane and are involved in identifying target cells for the HIV to infect (see Figures 2 and 3). The HIV surface proteins are also sites that the immune system can sometimes detect, making it possible to destroy that HIV virus particle (see Figure 4). For the dataset you will be working with, the researchers identified different forms, or clones, of HIV based on differences in the nucleotide sequence of a short, 285 base pair, region of gp120 called V3 (see Figure 1). The V3 region is known to be highly variable and involved with both host cell and antibody recognition. Characterizing the population of HIV in a subject based on the different versions of the V3 region sequence gave researchers a measure of HIV evolution that has potential clinical significance.

Figure 2: A schematic of the HIV virus. Note the positioning of the *env* gene products gp41 and gp120 on the surface of the particle.

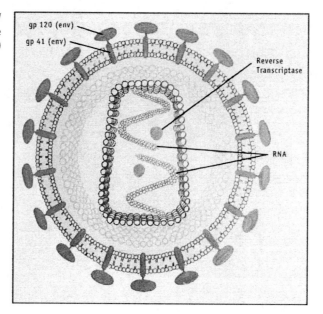

Figure 3: Schematic showing the role of gp120 in recognition of host cells.

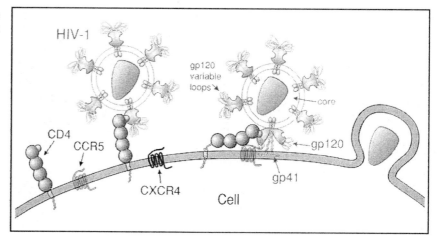

Activity 1: Looking at the NCBI Resources and HIV sequence data

As previously mentioned, these data were originally published as part of a research study looking at HIV evolution in different subjects. To learn a little more about the study and the data itself, this first activity involves searching the National Center for Biotechnology Information (NCBI) databases for additional information related to this research.

Part 1: PubMed

The NCBI provides access to a variety of databases including PubMed (a literature database) and GenBank (a nucleic acid sequence database). This federally funded research resource is very useful in part because the databases are linked together. This means that finding information in one area will allow you to look up related information in other areas. You should find the PubMed record for the Markham et al. article and then follow the links option for that record to find all the nucleotide sequences associated with that article.

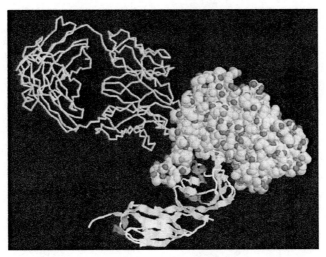

Figure 4: A representation of the gp120 protein structure (CPK space fill, right) and its interactions with the CD4 protein (ribbon, bottom) and an antibody (backbone trace, upper right). Structure published by Kwong et al., 1998.

NCBI website URL: http://www.ncbi.nlm.nih.gov

Original paper: Markham, R. B., W. C. Wang, A. E. Weisstein, Z. Wang, A. Munoz, A. Templeton, J. Margolick, D. Vlahov, T. Quinn, H. Farzadegan, X. F. Yu (1998). Patterns of HIV-1 evolution in individuals with differing rates of CD4 T cell decline. *Proceedings of the National Academy of Sciences* 95(21):12568-73. Pub Med ID: 98445411

- How did you search for the PubMed entry?

- What other ways might you have searched?

- What other types of related information are available?

Part 2: GenBank

In this section you will take a closer look at a GenBank record and the type of data that is stored there. Once you reach the nucleic acid data associated with the Markham et al. paper you will see that there are a variety of different ways to view the data.

The data you will be working with is coded to help you recognize its source. While all of the data are HIV sequences, each sequence is identified based on the subject it was taken from, the visit during which it was collected, and its clone number. Thus, each sequence has a code like S4V2-4 that can be read as subject 4, visit 2, 4th clone. Each clone is a unique sequence collected during a particular visit. Over 600 different HIV sequences were identified in these 15 subjects and published electronically in GenBank. Choose one of the GenBank records and view both the full record and the FASTA formatted sequence.

- What was the accession number of the sequence you chose?

- Which subject of the study was that HIV sequence from? Which section of the record contains information about who the HIV was collected from?

- Download several (4 to 6) sequences in FASTA format to your local hard drive by selecting several at the same time in the summary view so they are saved into a single text file. Be careful to remember where you put the file and what you name it so that you can find it later.

- Open the file that you saved with a word processor to confirm that you have the sequences and that they are in the FASTA format. In the FASTA format each sequence is preceded by a label which begins with the greater than sign (>).

Part 3: Introduction to the *Biology Workbench*

In order to analyze sequence data we will use the *Biology Workbench*, an on-line suite of bioinformatics tools. This section contains a brief introduction to the *Biology Workbench* that is intended to get you up and running quickly. There is more information about how the *Biology Workbench* is organized and how to use various tools in the "Orientation to the *Biology Workbench*" supplement on the *Microbes Count!* web site.

- Log in to the *Biology Workbench*: <http://workbench.sdsc.edu>.

- If you do not already have an account, you will need to set one up by following the Set up an account link.

- Once you have logged in, scroll down until you see the 5 buttons that take you to the different tool sets. This exercise uses nucleic sequence data, so follow the appropriate link.

- You should now see a scrolling list of tools for working with nucleic sequence data. Select Add new sequences and press the Run button. The next window allows you to enter sequences in a variety of ways. You can type them in directly, paste them in from another file, or upload a text file containing sequence data.

- Choose the Browse button to select the file you saved earlier from NCBI. Once you have selected the file use the Upload button to open it and read the data into this page.

- Once the labels and sequences appear in the data fields, choose Save to import that data into your *Biology Workbench* session.

- Each sequence should now appear as a data line below the list of analysis tools. Select one sequence and use the command View the Sequence to confirm that the sequence was successfully imported.

Now that you have been introduced to the *Biology Workbench* interface and procedures for uploading data files, you are ready to use the same procedures to upload a real dataset for analysis in the next activity.

- Look in the list of nucleic acid tools and find the ClustalW tool. Highlight the tool and then select Help to see more information about what this tool does.

- Select all of your sequences using the appropriate command and run a multiple sequence alignment using ClustalW.

- Look over the output and see if you can relate the differences in the sequences to the topology of the unrooted tree diagram and the pairwise similarity scores.

Note that the sequence labels on your tree will be the accession numbers for those sequence records. With just the accession numbers to describe the sequences, it may be difficult to think about the biological basis for the comparisons you just performed. Given what you have learned in Activity 1 you should be able to go back to GenBank and find additional information about the sequences you selected. We will use the ClustalW tool extensively in Activity 2. Look over this practice tree and think about how you will interpret your experimental trees to draw biological inferences from them.

- Go to the Session Tools and create a new session to store the sequences you will be working with in Activity 2.

Activity 2: Looking at the sources of HIV across subjects

For this activity you will work with HIV sequence data collected from 15 individuals from an injection drug using population in Baltimore. The goal of the study will be to determine if the HIV isolated from particular subgroups of subjects derives from a common source. In order to approach this question you will need to characterize the populations of HIV within an individual and quantify the differences between individuals. The work will be distributed among groups to make the research more manageable. You can then combine your own results with those from other groups to try to draw some conclusions about the entire population of subjects.

Most of your analyses will involve building multiple sequence alignments and distance-based unrooted trees. A multiple sequence alignment involves lining up several sequences so that comparisons can be made across sequences for each nucleotide position. Differences between related sequences can then be interpreted as nucleotide substitutions. Building alignments sometimes involves inserting gaps that represent areas where an insertion or deletion has taken place in one or more of the sequences. You will be using the ClustalW tool to build your alignments and distance-based unrooted trees.

Part 1: Looking at clustering across subjects

Upload the "visit_1_S1_S9.txt" and "visit_1_S10_S15.txt" files into the nucleic acid tool set of the *Biology Workbench* session you created in the last exercise. There are 2 data files because the *Biology Workbench* only allows up to 64 sequences to be uploaded at a time. Together these 2 files contain 97 sequences from the 15 subjects' first visits (see Table 1). The sequence labels for these data include information about the subject, visit and clone number for each sequence.

This research was a prospective study, meaning that the subjects were receiving regular HIV screening and the first visit for this dataset indicates the first time they tested HIV positive. Remember, each clone consists of one or more copies of the virus representing a unique sequence in the V3 region of the *env* gene.

Table 1: Summary of the number of different HIV sequences present early in the course of HIV for each patient.

Subject	Number of clones at visit 1
1	13
2	6
3	4
4	3
5	8
6	3
7	10
8	5
9	5
10	7
11	7
12	4
13	4
14	6
15	12

- As a preliminary analysis you should generate a multiple sequence alignment and distance tree for 12 of these sequences (3 clones from each of 4 subjects).

- Use the data table below (Table 2) to keep a record of the data you analyzed.

Table 2

Subject	Clone #

Interpreting unrooted trees. Unrooted trees represent the genetic distance between pairs of sequences. The trees are made up of branches (the lines between nodes) and nodes (the places where branches join or end). The ends of branches are called terminal nodes and represent sequences that you input into the analysis.

In tracing the branches between any two terminal nodes, or sequences, the total length traced reflects the genetic distance between those sequences. We can use the genetic distances as a first-order approximation of the evolutionary relationships between the sequences. The greater the genetic distance, the longer the time since they shared a common ancestor. However, because it is an unrooted tree we can't make any inferences about the direction of the evolutionary change. An important feature of unrooted trees are long internal branches. Long internal branches (i.e., branches that don't connect to terminal nodes) separate clusters of sequences. For example, Figure 5a suggests that sequences 1 and 2 are much more similar to each other than either of them is to sequences 3 or 4. On the other hand, Figure 5b does not provide strong evidence for grouping sequences.

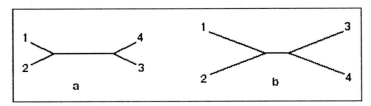

Figure 5: Hypothetical unrooted distance trees showing the difference between long (a) and short (b) internal branch lengths.

- Sketch your distance tree on paper so that you can make notes on it.

- Do the clones from each subject cluster together?

- Do some subjects' clones show more diversity than others?

- Do some of the subjects cluster together?

- Write a brief description of your tree and how you interpret the clustering pattern with respect to the similarities and potential evolutionary relationships between subjects' HIV sequences.

- Copy and paste your tree from the *Biology Workbench* into a word processor and print it out to share with the class.

Part 2: Quantifying diversity within and between subjects

This section will introduce several additional ways to quantify sequence similarity and difference. These measures will help you design and interpret your research in the next section.

Table 1 illustrates that different subjects have different numbers of clones in their initial samples. This is one crude way to quantify the diversity of the HIV within that individual at that time, assuming that each subject's sample contains approximately the same total number of viruses. The S statistic can also be used to quantify the diversity of sequence in a population. The value of S is simply the number of positions that vary, or are not identical, across all the sequences in an alignment.

- Select all the clones from a single visit of one subject and align them. From the alignment calculate S by counting the number of positions where there is at least one nucleotide difference across the collection of clones. Enter your data into Table 3 below.

- Import your alignment so that you can use it in a later section.

- Run the same analysis for a second and third subject and record your results in Table 3.

Subject	Number of Clones	S	Theta	Min difference	Max difference

Table 3

Next you will calculate θ (theta), which is an estimate of the average pairwise genetic distance. It is an estimate because it does not take into account the frequency of the different clones in the sample. θ is given by the formula

$$\theta = S / \left(\sum_{i=1}^{n-1} \frac{1}{i} \right),$$

where S is the statistic you just calculated above and the denominator of θ is a partial harmonic sum. For a sample of N = 5 the denominator would be calculated by $1 + 1/2 + 1/3 + 1/4 = 25/12$. The harmonic sum is used because we do not expect the number of differences to increase linearly with the number of sequences analyzed.

- Calculate θ for the three visits you chose to work with and enter the results in the datatable.

The last measures we will investigate are the minimum and maximum differences between any pair of sequences in an alignment. While S and θ provide some measure of difference across a set of sequences, the minimum and maximum pairwise distances look only at the extremes in terms of similarity and difference. You could obtain these values by looking directly at parts of larger alignments but it is easier to let the computer provide the pairwise comparisons for you.

- Use the Clustdist tool in the alignment tool set to generate a distance matrix for an alignment you saved.

- Select the highest and lowest pairwise scores and convert that percentage difference score into the raw number of differences by multiplying by the length of the sequence (285 in most cases).

- Round to the nearest integer, record your results in Table 3 and repeat the analysis for the other 2 subjects.

This measure can also be used to look at maximum and minimum differences across subjects.

- First, create a new alignment with all of the sequences from 2 subjects.

- Next, use Clustdist to generate a pairwise distance matrix for the alignment across subjects and find the minimum and maximum differences between the subjects sequences. Remember, you want to look only at pairwise distances across subjects, not within one subject.

- Record your data in Table 4 and repeat the analysis for the other pairs of subjects.

Pair of subjects being compared	Min difference	Max difference

Table 4.

- Post your results from this preliminary analysis so that other research groups can review your work as they design the research project in Activity 3.

Part 3: Design an analysis for looking at the sources of HIV

You now have the basic tools to compare and quantify groups of sequences. Look over the class results from the different groups' preliminary analyses and generate a list of possible specific questions to pursue. Design a small research project that will address one of your specific questions and contribute to the class-wide knowledge about the possible sources of HIV in this population.

- Prepare a report that communicates the question or idea you were testing, how you approached it with data collection and analysis, a summary of your findings, and an interpretation of your results.

Activity 3: HIV evolution research project

The data you used in Activity 2 was just a fraction of the data available from this study. The Markham et al. research followed this populations of 15 subjects over time and continued to sample and characterize their HIV populations at 6-month intervals for up to four years. In addition to studying the HIV population, the researchers estimated the immunological function of the subjects at each visit by measuring the number of CD4 T-cells in their blood samples. The CD4 T-cell count is an indicator of the health of the immune system because it represents a class of immune cells that are attacked by the HIV virus. A CD4 count below 200 cells/microliter is one of the diagnostic indicators used for defining Acquired Immune Deficiency Syndrome (AIDS).

There is a table summarizing the entire dataset on the web site. There are also files containing all the data from all the visits made by each subject. These data files are saved in FASTA format with labels like those used in the dataset you used in Activity 2.

- Print and review the data summary table for the complete Markham et al. data set.

- Discuss with your group any patterns you see in the summary data table and possible research questions related to those patterns.

- Design a small research study and prepare a scientific poster to share your project with the class.

Web Resources Used in this Activity

Biology Workbench (http://workbench.sdsc.edu)

The *Biology Workbench* was originally developed by the Computational Biology Group at the National Center for Supercomputing Applications at the University of Illinois at Urbana-Champaign. Ongoing development of version 3.2 is occurring at the San Diego Supercomputer Center, at the University of California, San Diego. The development was and is directed by Professor Shankar Subramaniam.

Platform Compatibility: Requires an internet connection and a current browser.

Additional Resources

Available on the *Microbes Count!* web site at http://bioquest.org/microbescount

Text

A PDF copy of this activity, formatted for printing

Data files from the Markham et al. (1998) research project

"Orientation to the *Biology Workbench*"

Related *Microbes Count!* Activities

Chapter 2: Searching for Amylase

Chapter 4: Molecular Forensics

Unseen Life on Earth Telecourse

Coordinates with Video IV: Reading the Code of Life

Relevant Textbook Keywords

Antibody, CD4 cell, Evolution, Gene, HIV, Mutation, Phylogenetic relationships, Retrovirus

Related Web Sites (accessed on 4/25/03)

Cells Alive HIV Tutorial
http://www.cellsalive.com/hiv0.htm

Centers for Disease Control and Prevention
http://www.cdc.gov/hiv/dhap.htm

Good info on HIV lifecycle
http://www.thebody.com/niaid/hiv_lifecycle/virpage.html

HIV Biology Background
http://medlib.med.utah.edu/WebPath/TUTORIAL/AIDS/HIV.html

Los Alamos HIV Lab
http://hiv-web.lanl.gov/

Microbes Count! Website
http://bioquest.org/microbescount

Unseen Life on Earth: A Telecourse
http://www.microbeworld.org/htm/mam/is_telecourse.htm

References (accessed on 4/25/03)

Mansky, L. M. and H. M. Temin (1995). Lower in vivo mutation rate of human immunodeficiency virus type 1 than predicted by the fidelity of purified reverse transcriptase. *Journal of Virology*. 69:5087-5094. (Pub Med ID: 7541846)

Markham, R. B., W. C. Wang, A. E. Weisstein, Z. Wang, A. Munoz, A. Templeton, J. Margolick, D. Vlahov, T. Quinn, A. Farzadegan, X. F. Yu (1998). Patterns of HIV-1 evolution in individuals with differing rates of CD4 T cell decline. *Proc. Natl. Acad. Sci.* 95(21):12568-73. (Pub Med ID: 98445411)

Kwong, P. D., R. Wyatt, J. Robinson, R. W. Sweet, J. Sodroski, W. A. Hendrickson (1998). Structure of an HIV gp120 envelope glycoprotein in complex with the CD4 receptor and a neutralizing human antibody. *Nature* 393(6686):648-59. (Pub Med ID: 9641677)

Figure and Table References

Figure 1.   http://hiv-web.lanl.gov/content/hiv-db/MAP/Env.html
             http://hiv-web.lanl.gov/content/hiv-db/MAP/landmark.html

Figure 2.   Courtesy of Sam Donovan

Figure 3.   http://www.niaid.nih.gov/daids/dtpdb/graphics/cellbin.gif

Figure 4.   Courtesy of Sam Donovan

Figure 5.   Courtesy of Sam Donovan

Chapter 5

Activities for Video V: Genetic Transfer

Microbial populations adapt quickly to new environments. This is largely due to the remarkable, yet routine transfer of genetic materials between microbes. The mechanisms of exchange have been explored and biotechnologists have utilized their knowledge of plasmids and viral transfer of nucleic acid information in numerous applications.

In this unit we can:

- select an a question to explore on the widespread use of genetically engineered Bt corn with its bacterial genes to protect against an exotic infestation,
- investigate the genetic basis for phenotypic similarities in bacterial strains by simulating cross-feeding, and
- examine conjugant populations from two strains of bacteria incubated for different periods of time as a strategy for mapping traits.

The Farmer and the Gene: A Case Approach to Bt Corn

This case study provides an opportunity to identify and then investigate questions about a real world application of genetic engineering. A look at changing agricultural control methods for the European Corn Borer sets the stage for exploring the uses of Bacillus thuringiensis since the 1930s.

Complements Please! No Compliments Necessary

Are mutant strains of bacteria that have the same nutritional phenotype identical? In the virtual MicroGCK lab, you can select colonies for further analysis to determine the underlying genetic structure of this similarity. This includes running a cross feeding complementation simulation for strains of nutritional auxotrophs.

Conjugation and Genetic Mapping

We know that bacteria share traits through conjugation. Can the frequency of specific exchanges help us map these traits? Analysis of the data from a series of MicroGCK conjugation simulations provides a method for genetic mapping.

The Farmer and the Gene: A Case Approach to Bt Corn

Ethel D. Stanley and Margaret A. Waterman

Video V: Genetic Transfer

Investigative case-based learning is a collaborative process in which individuals work together to identify and understand the scientific issues involved in the case at hand. You are encouraged to work in a group as you consider *The Farmer and the Gene.* Begin by asking someone to read the case below. (Note: The case is in the form of a short interview. Several of the figures in this activity offer additional information you may wish to consider as well.)

A case analysis worksheet structures the process of considering carefully the microbiological information your group already knows and what your group wants to know more about. You will be asked to generate your own questions and to pose specific problems for study. Then, choose one of the problems that interests the group.

You will need to decide which tools or resources to use. There are a number of resources provided in this activity and in the related web sites. On the *Microbes Count!* web site, there is an additional resource you may wish to consider, a file containing Protein Data Base information for five different Bt endotoxins. Your group must decide on a methodology for investigating the problem.

You should collect data and present your conclusions about the problem. Make sure the conclusions are supported by your findings. The manner of presentation can be negotiated with your instructor. Perhaps you will do a full lab report, or you could generate a presentation to another group, write a brochure on the topic for a specific audience, create a poster display, submit information to a local legislator or planning board, etc.

Figure 1. Control of the European corn borer in the United States circa 1920.

The Farmer and the Gene: An Interview

"Did you ever have to burn the corn fields to get rid of corn borers?"

"No, but my grandparents probably did. My Dad thought corn borers were a problem. I remember him checking the fields pretty regularly and spraying our corn with Bt–some kind of bacteria that killed the corn borers–at least a couple of times every summer. Now we just plant Bt corn."

"Bt corn? Are the bacteria mixed in when you plant?"

"It's corn that has a gene from the bacteria. I don't know why, but the bacteria make a protein that kills the corn borer and other insects."

"Do you think using Bt corn is safe?"

"I do. In fact, this year I'm planting a new hybrid that controls for rootworms as well as corn borers. Some folks think this is asking for trouble down the road. With the plants making Bt proteins in the field all season, the corn borers could get used to it. Some growers in the next county got mixed up in that StarLink fiasco, so they're really avoiding all Bt corn."

Figure 2. In the United States, Bt corn hybrids were grown on over 25 million acres in 2002.

Case Analysis Worksheet

(A copy of this worksheet is available on the *Microbes Count!* web site.)

1. Recognize potential issues.

List terms or phrases that seem to be important for understanding what the case is about.

2. Brainstorm for connections. Briefly discuss the following with the group.

What is this case about?

What are its major themes?

Keep track of major issues and questions that arise with the Know/Need To Know chart.

What do we already know?	What do we still need to know?

Identify *one* question or issue from the "need to know" list that your group wants to explore.

3. Obtain additional references or resources to help answer or explore questions.

These may include print resources, informational articles, data sets, results of simulations, maps, interviews, etc.

List *four* different resources you think would be important to use to learn more.

4. Design and conduct scientific investigations relevant to the question.

Investigations could be laboratory, field or computer-based experiences that the instructor arranges for the entire class or may be entirely student generated. Describe your plans.

5. Produce materials that support understanding of the conclusions.

These products can take many forms, from traditional papers and scientific reports, to posters, videos, pamphlets, consulting reports, role playing, interviews, etc.

What sorts of products might be produced as a result of investigations of the questions identified here?

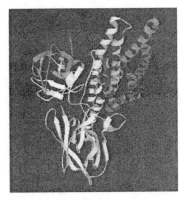

Figure 3. The structure of Cry3Bb, an insecticidal Bt protein that attaches to the lining of the insect gut and forms channels. This protein is effective against corn rootworms as well as corn borers.

Figure 4. The increasing use of Bt corn hybrids raises concerns. In the left image, note the proximity of the milkweed plant (preferred food of the Monarch butterfly) in the foreground to the corn in the background. In the right image, the mature tassels distribute pollen containing the Bt toxin over a large area.

Additional Resources

Available on the *Microbes Count!* web site at http://bioquest.org/microbescount

Text

A PDF copy of this activity, formatted for printing

Case Analysis Worksheet

Optional Assignment: Extending the Bt Corn Case

PDB Summary of Bt Endotoxins

Student Notes for Using Cases

Related *Microbes Count!* Activities

Chapter 9:   Mold Fights Back: A Challenge for Sanitation

Chapter 9:   Citrus Canker: Alternatives for Control

Chapter 9:   A Plague on Both Houses: Modeling Viral Infection to Control a Pest Outbreak

Chapter 10: Controlling Potato Blight: Past, Present and Future

Unseen Life on Earth Telecourse

Coordinates with Video V: Genetic Transfer

Relevant Textbook Keywords

Bacillus thuringiensis (Bt), Endotoxin, Hybrids, Stacked genes

Related Web Sites (accessed on 1/20/03)

Biology of the European Corn Borer
http://www.bio.org/food&ag/ncfap/corn.htm

Bt Corn Insect Refuge Calculator
http://profitablefarming.com/WebServices/BtCorn/InsectRefugeCalculator.asp

Bt crops suspected to hurt soil ecology
http://www.psrast.org/btsoilecol.htm

Bt toxin resources
http://www.nalusda.gov:80/bic/BTTOX/bttoxin.htm

Coming to terms with GM food, links to BBC resources
http://www.checkbiotech.org/root/index.cfm?fuseaction=briefings&keyword_id=3545

Corn Growers Fear Losses
http://bioquest.org/llsummer00/localnws.html

Data on expression levels of Cry3Bb
http://www.ucsusa.org/food/Cry3Bb_tables.pdf

Engineer a Crop (From PBS/Nova)
http://www.pbs.org/wgbh/harvest/engineer/

Ethics and GMO's – quite comprehensive outline
http://www.public.iastate.edu/~cfford/342EthicsandGMOs.htm

EPA registration material for Cry 1Ab Bt toxin
http://www.epa.gov/pesticides/biopesticides/factsheets/fs006444t.htm

European corn borer damage in Bt corn
http://www.ent.iastate.edu/imagegal/plantpath/corn/ecb/bteardam.html

Extending the Bt Corn Case
http://www.bioquest.org/lifelines/fract.html

Glossary of Biotech terms
http://www.ncbiotech.org/biotech101/glossary.cfm

Guess what's coming to dinner? Fun way to see impact of GE on food
http://www.pbs.org/wgbh/harvest/coming/coming.html

History of Bt
http://www.bt.ucsd.edu/bt_history.html

Harmony between agriculture and environment
http://www.ers.usda.gov/Emphases/Harmony/issues/genengcrops/terms.htm

Introduction: The European Corn Borer
http://www.ent.iastate.edu/pest/cornborer/intro/intro.html

LifeLines OnLine
http://bioquest.org/lifelines/index.html

Look What's Out There: March 2003
http://www.wvu.edu/~agexten/lookwhat/lwot303.pdf

Microbial mutation clock
http://www.pbs.org/wgbh/evolution/survival/clock/

Microbes Count! Website
http://bioquest.org/microbescount

Pesticide use likely to feel bite of new biotech corn (June 2002) Bt for
rootworm
http://www.ohio.com/mld/beaconjournal/news/nation/3616259.htm

Pros and Cons of Genetically Engineered crops
http://www.wvu.edu/~agexten/ipm/animals/genetic2.htm

Refuges for Resistance Management: Bt Corn and European Corn Borers
http://spectre.ag.uiuc.edu/cespubs/pest/articles/199904d.html

Should I plant Bt corn?
http://msucares.com/crops/corn/corn13.html

Student Notes for Using Cases
http://www.bioquest.org/snotes2.html

Union of Concerned Scientists comments to EPA about new Cry3Bb for corn
rootworm: Detailed; discusses also antibiotic resistance markers used in
genetic engineering
http://www.ucsusa.org/food_and_environment/biotechnology/page.cfm?pageID=1014

Unseen Life on Earth: A Telecourse
http://www.microbeworld.org/htm/mam/is_telecourse.htm

US Distribution of Bt corn
http://www.ncga.com/biotechnology/BtMaps/USMaps1.htm

References

Stanley, E. D. and M. A. Waterman (2000). LifeLines OnLine: Curriculum and Teaching Strategies for Adult Learners. *Journal of College Science Teaching,* March/April 2000, 306-310.

Waterman, M.A. (1998). Investigative case study approach for biology learning. *Bioscene: Journal of College Biology Teaching* 24(1): 3-10.

Figure and Table References

Figure 1.   Caffrey, D. J. (1919). The European corn borer. Farmers' Bulletin 1046. United States. Dept. of Agriculture

Figure 2.   Courtesy Ethel Stanley

Figure 3.   Modified images from the Protein Data Bank web site http://www.rcsb.org/pdb/

Figure 4.   Courtesy Ethel Stanley

Complements Please! No Compliments Necessary

John R. Jungck and John N. Calley

Video V: Genetic Transfer

Phenotypes, Genotypes, and Metabolism

If two bacterial colonies on a Petri plate require the same combination of nutrients to grow, does this mean that they are genetically identical?

If these mutant colonies are genetically different, can we determine this just by looking more closely at their metabolism?

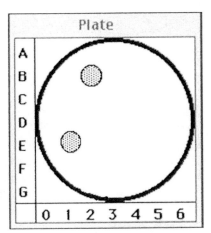

Figure 1. This screenshot from the *Microbial Genetics Construction Kit* shows two colonies growing on an incomplete nutrient medium that failed to support other field plate colonies.

If you want to learn about the characteristics of a diverse collection of bacterial colonies, you can go into the microbiology lab and use the techniques of replica plating and auxonography to determine their nutritional phenotypes. (In Chapter 1 of this book you can perform these experiments in a virtual lab using the *Microbial Genetics Construction Kit (µGCK)*. See the "How Do You Know a Microbe When You See One?" activity.) However, you have not actually investigated the genetics of these bacteria yet. Even after determining the nutritional phenotypes of a collection of bacterial colonies (see the phenotypic matrix generated by *µGCK* in Figure 2 for an example) you cannot assume that you know anything about the genotypes that underlie these phenotypes.

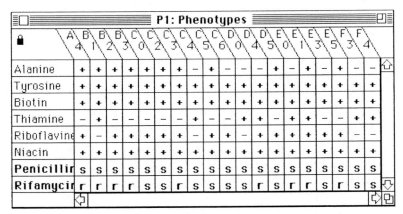

P1: Phenotypes

	A4	B1	B2	B3	C0	C2	C3	C4	C5	C6	D0	D4	D5	E0	E1	E3	E5	F3	F4
Alanine	+	+	+	+	+	+	+	−	+	−	−	−	+	−	+	−	+	−	−
Tyrosine	+	+	+	+	+	+	+	+	+	+	+	+	+	+	+	+	+	+	+
Biotin	+	+	+	+	+	+	+	+	+	+	+	+	+	+	+	+	+	+	+
Thiamine	−	+	−	−	−	−	−	+	−	−	+	+	−	+	+	−	−	+	+
Riboflavine	+	−	+	+	+	+	+	−	+	+	−	+	+	+	+	+	+	−	−
Niacin	+	+	+	+	+	+	+	+	+	+	+	+	+	+	+	+	+	+	+
Penicillin	s	s	s	s	s	s	s	s	s	s	s	s	s	s	s	s	s	s	s
Rifamycin	r	r	r	r	s	s	r	s	s	s	s	r	s	r	r	s	r	s	s

Figure 2. Completed phenotypic matrix in *µGCK*. Colonies are listed across the top and the nutrients used to test colonies for growth are listed on the left.

A plus (+) indicates that growth occurred without the addition of the nutrient under consideration—the colony is prototrophic, that is, it does <u>not</u> require the nutrient. A minus (-) indicates no growth—the colony is auxotrophic and <u>does</u> require the nutrient for growth to occur. Sensitivity (s) or resistance (r) to the antibiotics penicillin and rifamycin is also recorded in this example.

To determine the underlying genetic structure of two mutant strains of bacteria that have the same nutritional phenotype (that is, that require the same nutrients for growth) it is necessary to perform some additional tests.

Some Background

Throughout the first half of the twentieth century, the gene was described as the unit of mutation, recombination and function. After the Watson-Crick-Franklin discovery of the double helical nature of DNA structure, Seymour Benzer reasoned that if a string of nucleotides corresponded to a gene, then a gene should be redefined. No longer could the gene be 'the' unit of mutation, recombination, and function. Benzer also sought to clarify the underlying correlation between structure and function implied by this new view of a gene.

Mutation: A mutation could be as small as substituting, deleting, or adding a single nucleotide base pair; it did not have to involve a complete gene. Benzer coined the word "muton" to correspond with the smallest unit of mutation.

Recombination: Similarly, recombination could occur within genes as well as between genes; as close as between two successive nucleotide base pairs. Benzer coined the word "recon" to correspond with the smallest unit of recombination.

Function: Benzer realized that just because two mutants were in the same gene and affected the same function, he still did not know if the two mutants were parts of a contiguous string of nucleotides that encoded the same functional product. It was well known that hemoglobin is a protein with multiple subunits (alpha and beta in most adult humans). Benzer reasoned that if two adjacent contiguous strings of nucleotides on the same chromosome encoded different subunits of one protein complex such as hemoglobin, then it should be possible to construct a test to differentiate between them. In the late 1930's Nobel Laureate E. B. Lewis developed such tests using fruit flies. However, it was the extensive work of Beadle and Tatum with bread mold and Pontecorvo with yeast that popularized the standard tests for function. Their methods were designed to determine whether two mutations represented alternative alleles of just one gene or if they affected two genes producing different functional products.

The Cis-Trans Complementation Test for Allelism and Function

The classic methodology for determining whether two independent mutations are in the same gene (or functional unit) or in different genes is called the Cis-Trans Complementation Test.

Consider the situation where there are two mutations in the piece of DNA that codes for a particular functional product, say the end-product in a metabolic pathway. Each mutation prevents the synthesis of a different critical intermediate product in the metabolic pathway. Without both of the intermediate products, no end product will be produced and no growth will occur.

/-   -/
/+   +/

Cis configuration of two mutants in a diploid pair of homologous chromosomes

If the two mutations are on the same gene (functional unit), then that gene cannot synthesize either of the intermediate products and the metabolic end-product is not produced. This is called the cis configuration.

/+ / -/
/- / +/

Trans configuration of two mutants in a diploid pair of homologous chromosomes

If, on the other hand, the mutations are on different functional units (the trans configuration), each gene can produce its intermediate product, and the end-product of the pathway can be synthesized. In the latter case, the mutations are said to complement each other.

Complementation: Cross-feeding Tests for Allelism and Function

Complementation is a powerful tool for examining the genetic basis for differences among mutants with similar phenotypes. However, the classic cis-trans complementation test for determining whether two independent mutations are in the same gene or in different genes is limited to recessive mutations in diploid organisms. Furthermore, this technique does not generate a qualitative (yes or no) result, but instead gives a quantitative result because of potential confusion with recombination. To avoid these complications, Norton Zinder developed a complementation test that utilized a technique called cross-feeding to explore genetics in microbial populations. This technique does not require the construction of diploids and produces results that cannot be confused with recombination since no DNA is exchanged between strains.

The cross-feeding test uses mutant strains of auxotrophic bacteria. While phenotypically identical, each complementing mutant is unable to synthesize one of the precursors to a terminal product required for growth; no growth will occur unless the needed nutrient is supplied in the growth medium. Each auxotroph used in the test requires the addition of a different nutrient. Furthermore, the microbes used will normally produce the precursor metabolites in such excess that they can be found in the extracellular medium surrounding the bacterial colonies.

In Zinder's technique two different auxotrophic strains are placed in opposite sides of a U-shaped tube with a selectively permeable glass filter between them (see Figure 3). The holes in the filter allow small molecules such as amino acids and vitamins to freely pass back and forth, but not macromolecules like DNA,

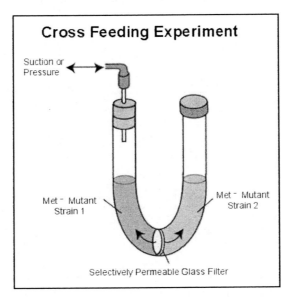

Cross Feeding Experiment

Suction or Pressure

Met⁻ Mutant Strain 1

Met⁻ Mutant Strain 2

Selectively Permeable Glass Filter

Figure 3. Apparatus for the Cross-Feeding Complementation Test.

RNA, protein, or starch. If the first strain produces an intermediate metabolite that the second strain cannot produce, the second strain can still grow if the "missing" metabolite passes through the filter and is available for utilization in the synthesis of the terminal product. Similarly, if the second strain produces enough of its product, the product will pass back through the barrier and be available to the first strain, which will now be able to grow. This method is known as cross-feeding and it is the method simulated in the *Microbial Genetics Construction Kit (µGCK)* application.

Using the Cross-Feeding Complementation Test to Investigate Multiple Complementation Groups

In this activity you will use *µGCK* to investigate a simulated system with multiple complementation groups (see the exercise below).

Basic Steps for a Cross-Feeding Complementation Test in *µGCK*.

- Use the Phenotype Worksheet to choose two or more colonies that are auxotrophic for the same nutrient. If you want to see whether there is more than one complementation group, you can choose all of the colonies at once.

- Use the Media Matrix to create a new medium that does not contain the nutrient for which your colonies are auxotrophic.

- Innoculate the tube containing the new medium with your colonies and observe to see if growth occurs. You may want to plate out the tube if it is difficult to tell if growth has occurred in the tube.

- Record your results for each nutrient that you test in a Complementation Matrix worksheet (the complementation relationships will be different for every nutrient that you test). You can drag rows and columns of the Complementation worksheet to organize the information more conveniently. This can help to clarify complementation groups. In Figure 4, an unorganized and an organized complementation matrix are illustrated. The cells at the

Figure 4: (a) An uncompleted Tyrosine Complementation Matrix. (b) A completed and rearranged Leucine Complementation Matrix with five complementation groups. (Shading was added for emphasis.)

The mathematics associated with the symmetrical rearrangements and the connection of all minus signs to the main diagonal (Fig 4b) are elaborated in *BENZER*, a *BioQUEST Library* module. The software is named after Seymour Benzer, who used complementation and deletion mapping to explore the fine structure of the gene.

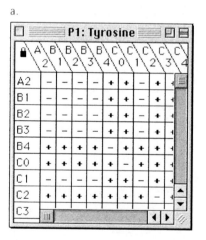

a.

P1: Tyrosine

	A2	B1	B2	B3	B4	C0	C1	C2	C3	C4
A2	-	-	-	-	+	+	-	+	+	
B1	-	-	-	-	+	+	-	+	+	
B2	-	-	-	-	+	+	-	+	+	
B3	-	-	-	-	+	+	-	+	+	
B4	+	+	+	+	-	+	+	+	+	
C0	+	+	+	+	+	-	+	+	+	
C1	-	-	-	-	+	+	-	+	+	
C2	+	+	+	+	+	+	+	-	+	
C3										

b.

P1: Leucine

	C4	C5	A2	B4	D4	F1	E3	F5	C2	G3	A3
C4	-	-	+	+	+	+	+	+	+	+	
C5	-	-	-	+	+	+	+	+	+	+	
A2	+	-	-	-	+	+	+	+	+	+	
B4	+	-	-	-	-	-	+	+	+	+	
D4	+	+	+	-	-	-	+	+	+	+	
F1	+	+	+	-	-	-	-	+	+	+	
E3	+	+	+	+	-	-	-	-	+	+	
F5	+	+	+	+	+	-	-	-	-	-	
C2	+	+	+	+	+	+	+	-	-	-	
G3	+	+	+	+	+	+	+	-	-	-	
A3											

intersections of rows and columns record whether or not the corresponding colonies complement each other. A plus (+) indicates complementation, a minus (-) indicates no complementation, and a question mark (?) indicates that the complementation status is unknown.

The number and members of complementation groups can be inferred from the reorganized Complementation Matrix. Furthermore, the type of mutation and the number of distinct polypeptides or enzyme-catalyzed metabolic steps can be inferred. Mutations that belong to two different complementation groups are probably deletions that span the junction of both cistrons rather than point mutations. This hypothesis could be tested in the lab by growing them up in bulk culture to test for reversibility; point mutations are often reversible and deletions usually are not reversible.

An installer and detailed instructions for using the *Microbial Genetics Construction Kit (µGCK)* are available on the *Microbes Count!* web site.

Exercise:

1. For your Field Plate, determine how many metabolites you can test for complementation.

2. Similarly, for your Field Plate, determine the colony members of each complementation group for each metabolite with multiple auxotrophs.

3. How many complementation groups are there for each metabolite?

4. For both of the above two tasks, illustrate your data in multiple visual forms that relate genotypes to observed data that you have collected.

5. Assign the members of each complementation group to steps in a plausible metabolic pathway for each metabolite.

6. Consult the literature on known biosynthetic pathways in bacteria for the substances in question. Determine some of the precursors and enzymes involved in known pathways and develop a strategy for how you would test which complementation groups correspond to genes for particular enzymes or cistrons for protomers of oligomeric enzymes.

   Besides your textbook, we suggest that you consult two online metabolic pathway databases:

   1. EcoCyc developed by Monica Riley and colleagues at the Marine Biology Laboratory in Woods Hole, Massachusetts and Peter D. Karp, at SRI International, Menlo Park, California.

        http://ecocyc.org/ecocyc/ecocyc.html

   2. The Kyoto Encyclopedia of Genes and Genomes (KEGG) developed by Minoru Kanehisa and colleagues of the Bioinformatics Center, Institute for Chemical Research, Kyoto University.

        http://www.genome.ad.jp/kegg/

   Another tool which you may find very helpful is PathDB: A second generation metabolic database developed by Blanchard, J.L., Bulmore, D.L., Farmer, A.D., Gonzales, M., Steadman, P.A., Waugh, M.E., Wlodek, S.T. and Mendes, P., National Center for Genome Resources, 2935 Rodeo Park Drive East, Santa Fe, NM 87505, USA.

     http://www.ncgr.org/software/pathdb ; email: pathways@ncgr.org

John R. Jungck and John N. Calley   The BioQUEST Curriculum Consortium

Software Used in this Activity

Microbial Genetics Construction Kit

John N. Calley (Eli Lilly and Company) and John R. Jungck (Beloit College)

Platform Compatibility: Macintosh only

Additional Resources

Available on the *Microbes Count!* web site at http://bioquest.org/microbescount

Software

Microbial Genetics Construction Kit

Text

A PDF copy of this activity, formatted for printing

"Complementation Testing Using the *Microbial Genetics Construction Kit*"

Related *Microbes Count!* Activities

Chapter 1: How Do You Know a Microbe When You Find One?

Chapter 5: Conjugation and Mapping

Unseen Life on Earth Telecourse

Coordinates with Video V: Genetic Transfer

Relevant Textbook Keywords

Antibiotics, Cistrons, Complementation, Cross-feeding, Genotype, Inoculum, Metabolic Pathways, One Gene–One Enzyme Hypothesis, Phenotype, Replica Plating, Serial Dilution

Related Web Sites (accessed on 2/20/03)

American Society for Microbiology
http://asmusa.org

EcoCyc. Riley, M. et al. at the Marine Biology Laboratory, Woods Hole, Massachusetts, and Karp, P. D., at SRI International, Menlo Park, California. http://ecocyc.org/ecocyc/ecocyc.html

The Kyoto Encyclopedia of Genes and Genomes (KEGG). Kanehisa, M. et al. Bioinformatics Center, Institute for Chemical Research, Kyoto University. http://www.genome.ad.jp/kegg/

Microbes Count! Website
http://bioquest.org/microbescount

PathDB: A second generation metabolic database. Blanchard, J. L., D. L. Bulmore, A. D. Farmer, M. Gonzales, P. A. Steadman, M. E. Waugh, S. T. Wlodek, and P. Mendes. National Center for Genome Resources. http://www.ncgr.org/software/pathdb

Related wet lab activity
   http://www.rickhershberger.com/bioactivesite/bio250/lab/Yeast_Complementation.pdf

Unseen Life on Earth: A Telecourse
   http://www.microbeworld.org/htm/mam/is_telecourse.htm

References

Adolph, K. W., editor. (1996). *Microbial genome methods*. CRC Press: Boca Raton.

Calley, J. N. and J. R. Jungck (2001). Microbial Genetics Construction Kit. In *The BioQUEST Library Volume VI*, Jungck, J. R. and V. G. Vaughan, Editors. Academic Press: San Diego, California.

Fincham, J. R. S. (1966). *Genetic complementation*. W.A. Benjamin: New York.

Jungck, J. R., V. Streif, I. Ceraj, and S. Everse (2001) *BENZER: An Interval Graph Tool for Deletion Mapping, Restriction Mapping, Complementation Mapping, Sequencing, and Food Web Analysis*. In *The BioQUEST Library Volume VI*, Jungck, J. R. and V. G. Vaughan, Editors. Academic Press: San Diego, California.

Maloy, S. R., J. Cronan Jr., D. Freifelder, and J. E. Cronan (1994). *Microbial Genetics*, 2nd Edition. Jones & Bartlett Publishers: Boston, MA.

Auxanographic Methods For Yeast Identification (1979) Videotape #V519
   http://www.state.in.us/isdh/labs/av_info.htm

Figure and Table References

Figure 1.   Screen shot from *Microbial Genetics Construction Kit*

Figure 2.   Screen shot from *Microbial Genetics Construction Kit*

Figure 3.   Joanna Cramer

Figure 4.   Screen shot from *Microbial Genetics Construction Kit*

Conjugation and Genetic Mapping

John R. Jungck and John N. Calley

Video V: Genetic Transfer

Bacterial Genetics

When studying bacterial mating

Lederberg found it frustrating

to make things look nice

and do everything twice

he invented replica plating.

H. Ibelgaufts (1997)

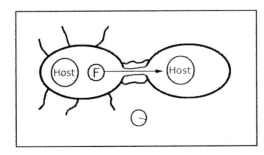

Figure 1. Conjugation between a donor Hfr bacterium and a F- strain bacterium. Note the pilus connecting the bacteria and the transfer of genetic material.

In two previous activities, the *Microbial Genetics Construction Kit (μGCK)* provided a virtual lab where phenotypic differences between bacterial strains were determined by replica plating and auxonography. Genotypic differences were investigated by complementation (cross-feeding) experiments. However, the characterization of a bacterial strain by auxonography and complementation still does not provide us with adequate information to infer where the genes (cistrons) are and in what order they occur. In this activity, we will use *μGCK* to explore questions about mapping genes.

How are genes arranged in a bacterial genome? Does this order relate to their metabolism? Do genes involved in antibiotic resistance move from one cell to another within a generation? Or do resistant individuals only arise by means of mutation and reproduction? These questions require us to consider what is known about the mechanics of genetic transfer in bacteria and to utilize new techniques to produce data needed for mapping genes.

Background

Bacteria are usually haploid; that is, they carry only one copy of each gene in their genome. However, some bacteria have multiple DNA molecules within their cells. Separation of the DNA molecules in *Agrobacterium tumefaciens* by pulsed field gel electrophoresis provides evidence for four different genomes. (Allardet-Servent et al. 1993) The four molecules differed in both size (ranging from 200,000 to three million nucleotides) and topology (three circular chromosomes or plasmids and one linear chromosome). Should only one be considered the bacterium's "real" genome or are all four together its genome?

While DNA transfer between bacteria had been demonstrated in 1928 via transformation experiments, bacterial "sex" (conjugation) was not inferred from experiments until the late 1940's. Lederberg and Tatum demonstrated in 1946 that physical contact was required for this transfer of genes (Figure 2.) When the two parental strains were separated by a sintered glass filter, no transfer occurred.

Figure 2: Lederberg and Tatum experiment. Neither parental strain could grow in the counter-selective medium. Only the presence of recombinant progeny produced growth.

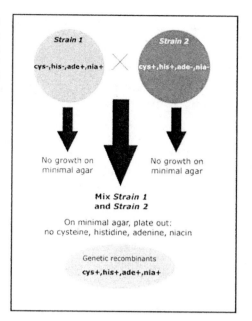

Later electron micrographs dramatically confirmed this need for contact by demonstrating the presence of a hollow tube (called a pilus) connecting the donor and recipient strains of bacteria (Figure 1.)

The pilus is composed of proteins called pilins which self assemble into hollow tubes. Pilin has a rare feature where the amino and carboxyl ends of the protein are covalently linked in a circle (Kalkum et al. 2002)

Although most textbook illustrations show DNA moving through a tube connecting the cells, it is not known whether DNA actually moves through pili or whether the pili simply bring members of different strains into close enough proximity for mating to occur.

In the 1950's, Lederberg and Lederberg produced genetic maps by investigating the relative frequencies of reciprocal crosses. Experiments were run with known genetic markers of the donor and recipient strains reversed. If three markers were chosen, the transfer of the central marker interrupted the transfer of its two distal neighbors. They reasoned that a single pair of double crossovers was needed to incorporate one long block containing both of the single markers whereas to incorporate two blocks containing each of the single markers would have required two pairs of double crossovers. The quantitiative difference expected was so great that the two experiments yielded qualitatively different results.

Jacob and Wollman (1961) found a more efficient and rapid method for determining the order of genes on a bacterial chromosome using interrupted conjugation, in which an Hfr donor strain was mixed with an F- auxotrophic strain. Since conjugation requires physical contact for the transfer of DNA from the donor to recipient cell, they used a blender to shear the cells from one another. The blender was turned off and on at specific intervals and then successive samples were plated on counter-selective media. The presence of markers on the media was used to calculate their relative position on the chromosome. See Figure 3, stages C and D.

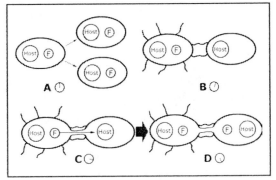

Figure 3. A. Replication of Hfr bacterium; B. Physical contact between donor Hfr and F- strain bacteria signals the start of conjugation; C. Pilus formation completed and transfer of genetic material from Hfr bacterium initiated, and; D. Transfer of genetic material complete. Note the time intervals.

While the order of entry is constant from one pair of cells to the next, not all of the cells are synchronous in their actual time of pilus formation, connection, and initiation of transfer. Thus, the markers continue to come across over a longer period of time for the full population being studied (see Figure 4). Therefore, identifying the first time that each marker starts to appear in the recipient strain is used to mark their position on the chromosome. The map uses minutes as the unit of measurement. (See Figure 5)

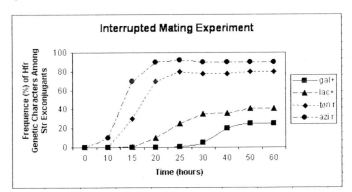

Figure 4. Graph of the kinetics of entry of markers.

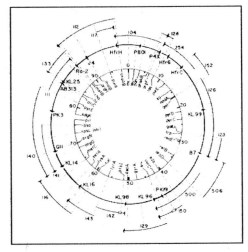

Figure 5. Map of *Escherichia coli* chromosome in minutes. Note that different strains have the F plasmid inserted at different spots (Hfr C, G, and H). Thus, while the order of markers is consistent, they are circularly permuted with respect to one another and these incongruities have to be resolved. In *μGCK*, only one Hfr strain is employed within one simulation.

Simulating Conjugation with *μGCK*

"A Tour of Conjugation Mapping" is provided on the *Microbes Count!* web site to familiarize you with the basics for using the simulation, but here is a brief description of how the model will work. All models are necessarily simplifications of the process being explored. The model of conjugation employed by *µGCK* is restricted to forms of conjugation mediated by the F plasmid in the Hfr configuration. In this module, there is only one insertion point for the F plasmid.

Once the user chooses to conjugate two populations of bacteria, one of which is Hfr and one of which is F-, the model must simulate the actual act of conjugation. The outcome of a conjugation cross is always described in terms of the number of resultant recombinants observable by conventional selective techniques. This is used to generate initial gross maps of a genome via time of entry mapping and may also be used to some extent for more fine structure mapping by looking at recombination frequencies.

An isogenic Hfr strain has one particular F insertion point so it always transfers the bacterial chromosome starting at a particular origin and in the same direction. If we look at just one mating pair, the chromosome is transferred from the Hfr donor to the F- recipient at a roughly constant rate (after a variable set-up time) until the connection is broken. This is accomplished by either deliberate shaking by the experimenter or at random with a constant probability per unit time. Once a segment of donor chromosome has entered the recipient cell it may recombine with the local chromosome. The actual recombination process is not well understood. Parts of the donor chromosome that do not recombine may persist for a while (possibly even long enough to recombine with the recipient's descendants), but eventually degrade.

Usually, of course, one is not dealing with a single conjugating pair but with a very large population of bacteria that meet and begin to conjugate over some time interval. In this model we have made a number of simplifying assumptions. First, the donor chromosome does not persist after recombining or failing to recombine with the initial recipient. Also, we assume that all pairs begin to conjugate at the same time, and that chromosome transfer always occurs at the same rate (roughly that of *E. coli* at 37°C on optimal media). Then, since recombination mapping using conjugation is not that common, we have made the drastic assumption that the probability of a crossover between any two beads is constant, independent of other nearby crossovers (see the Model Background section). This means that we cannot model interference. The computational complexity of the problem is thereby greatly reduced.

We calculate the proportions of various recombinants in the final population by calculating, for every cistron that differs between the parental types, the probability that it will enter a recipient cell in the time available, and the probability that it will recombine with the recipient chromosome. Since we have postulated that all recombination events are independent, we then find the probability of multiple recombinants by multiplying the probabilities for individual cistrons.

In more detail, the probability that a particular donor cistron will recombine with the recipient genome is stronly influenced by the length of donor chromosome on each side of the cistron. This calculation has to be re-done for each possible length of entering chromosome. For further information on how this model is constructed see the Model Background section.

Before beginning to use *μGCK* to do genetic mapping, it is highly recommended that you try some two- and three-factor bacterial mapping problems from your genetics, microbial genetics, or molecular text or a microbiological problem web site such as that of Professor Stanley Malloy: http://www.life.uiuc.edu/micro/316/316resources/problems/recombination/mapping/crosses/

Activities

1. Using the problem in the *μGCK* folder on the *Microbes Count!* web site labeled "Mapping Markers1," produce a map of the three auxotrophic markers. In this case, you are provided with a completely filled phenotype matrix and set of complementation matrices as well as the Hfr/F- status of each colony. Caution: make sure that you use counter-selective media so that neither parent strain grows. Double-check that you have actually produced genetically different strains by growing cells from a tube that was produced by conjugation and that has acquired a new phenotype. You can do this by plating an aliquot on a plate without the essential nutrients (note the difference from doing the same thing after a complementation experiment).

2. Using the problem in the *μGCK* folder on the *Microbes Count!* web site labeled "Mapping Markers2," produce a map of the three auxotrophic markers. In this case, you are not provided with a completely filled phenotype matrix and set of complementation matrices, but you still are given the Hfr/F- status of each colony. Observe the same cautions as before.

3. Class project: Using the problem in the *μGCK* folder on the *Microbes Count!* web site labeled "Mapping Markers3," produce a map of the six auxotrophic markers and all of their separate cistrons (as defined by their complementation groups). Again in this case, you are provided with a completely filled phenotype matrix and set of complementation matrices as well as the Hfr/F- status of each colony. Because this problem is much more complex than either of the first two problems, we suggest that each group of students map only two or three markers and their respective cistrons and then all groups compare gene positions, resolve discrepancies, and construct an overall map of all markers.

Model Background

The Species Model

μGCK models a bacterial chromosome as a string of beads, where each bead corresponds to some constant number of codons. The correspondence of codons to beads is set by the model's requirement (for implementation reasons), that a cistron consist of from 1 to 16 beads. If we accept the common rough approximation that a cistron has about 1000 base pairs or 350 codons (each codon is three base pairs long), and assume that this corresponds to 3.5 beads, then a bead can be thought of as roughly 100 codons. This means that the total range of cistron sizes available is from 100 to 1600 codons.

Bacterial chromosomes are often measured in minutes, where a minute is the length of chromosome transferred between two conjugating bacteria (at 37°C) in one minute. Conjugation in *E. Coli* occurs at about 1000 base pairs per second which means that we have about 200 beads per bacterial minute.

Cistrons are parts of groups which together form and regulate a biochemical pathway with potentially phenotypically observable results. For the purposes of this model, these groups of cistrons are referred to as markers. A marker may consist of from one to n cistrons where n is some maximum that must be set when *μGCK* is compiled (at the moment n= 5). For now, the cistrons comprising two different markers must be disjoint.

Markers in *μGCK* may be of three types: Resistance, Auxotrophic, or Degradative. Resistance markers control sensitivity and resistance to antibiotics (or possibly to other chemical or phage attack). Auxotrophic markers control a bacterium's ability to synthesize needed nutrients from the basic materials available in a minimal medium. Degradative markers (not completely implemented yet) control the ability to degrade complex sugars into basic glucose.

The relationship between a marker and its component cistrons is extremely simple at the moment. Each cistron is assumed to code for a product which takes part in a serial biochemical pathway. If any part of the pathway is blocked by a defective cistron, the marker is defective. Each cistron product may be marked as non-diffusible, which means that it cannot complement at all, diffusible, so that it can complement another defective cistron within the same cell (in a merozygote), or cell wall diffusible so that two different cells can complement in a cross feeding experiment. In the current *μGCK* all cistron products are cell wall diffusible.

Definitions

limit = Maximum length of chromosome that could possibly enter the recipient. This is a function of the time before agitation and the rate of chromosome transfer.

start = length of chromosome between the transfer origin and the beginning of the entering cistron.

b = probability of breakage between two entering chromosome beads.

c = probability of a crossover between two chromosome beads.

e(l) = probability that a particular length *l* of donor chromosome will enter the recipient.

$$e(l) \approx b(1-b)l$$

O(l) = probability of an odd number of crossovers in a chromosome segment of length *l* beads.

$$O(l) = \sum_{x=1}^{l} \binom{l}{x} c^x (1-c)^{l-x} \text{ for odd x}$$

r = probability that a cistron, once in the recipient, will yield a recombinant.

$$r = \sum_{x-start}^{limit} e(x)O(start)O(x - start)$$

The Model

The probability that a cistron will enter a donor and produce a recombinant is $e(start) \cdot r$.

Hayes (1968) mentions .06 as a reasonable value for b; Curtiss (1969) states that the correct value is .064. We use .06/200 = .0003 per bead.

We find a reasonable value for c from measures of linkage distances correlated with time-of-entry differences. Hayes suggests that two markers separated by 5 minutes will exhibit a recombination probability of 50%. Curtiss suggests a recombination rate of 20% per minute. We derived a value of about .05 per bead for c.

We make a number of simplifications to lower the computational load to manageable proportions. First, we use a Poisson approximation to the binomial described above (see the previous discussion of Poisson assumptions in the "Population Explosion: Modeling Phage Growth" activity in Chapter 1.) Second, we evaluate entry and recombination probabilities not at every bead but at every nth bead.

The needs of users of the μGCK conjugation model will vary. We anticipate that most of the use of conjugation will be for time-of-entry mapping. Since this does not require any analysis of recombination and since the calculation of recombination frequencies is quite time consuming, we have provided one model of conjugation which does no calculation of recombination at all. It simply finds the probability of entry for markers and assumes that any marker that enters a recipient cell will be incorporated into its genome. Another model does calculate recombination probabilities, but with a relatively large step size so that its accuracy over short distance is not to be trusted too far. This screen will be invisible to you if you only have the pre-prepared μGCK problems on *Microbes Count!* web site (the full version is only available as part of *The BioQUEST Library*). However, we thought that it was useful to show it here (Figure 6) so that users have some idea of how we constructed problems in the first place.

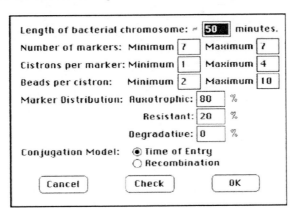

Figure 6. Genome Configuration editing.

John R. Jungck and John N. Calley

The BioQUEST Curriculum Consortium

Software Used in this Activity

Microbial Genetics Construction Kit

John N. Calley (Eli Lilly and Company) and John R. Jungck (Beloit College)

Platform Compatibility: Macintosh only

Additional Resources

Available on the *Microbes Count!* web site at http://bioquest.org/microbescount

Software

Microbial Genetics Construction Kit

The three pre-defined problems used in this activity: Mapping Markers1, Mapping Markers2, and Mapping Markers3

Text

A PDF copy of this activity, formatted for printing

"Complementation Testing Using the *Microbial Genetics Construction Kit*"

Related *Microbes Count!* Activities

Chapter 1: How Do You Know a Microbe When You Find One?

Chapter 4: Searching for the Hereditary Molecule

Chapter 5: Complements Please!

Unseen Life on Earth Telecourse

Coordinates with Video V: Genetic Transfer

Relevant Textbook Keywords

Complementation Groups, Conjugation, Mapping, Minutes, Pilus, Transformation

Related Web Sites (accessed on 2/20/03)

Birds Do It, Bees Do It: Lederberg and Tatum Discover Bacterial Mating and Phage Recombinaton (1946). Chris Evers, Access Excellence
http://www.accessexcellence.org/AB/BC/Birds_Bees.html

Joshua Lederberg: Bacterial Genetics, National Library of Medicine
http://www.nlm.nih.gov/hmd/lederberg/bacterial.html

Microbes Count! Website
http://bioquest.org/microbescount

Two- and three-factor crosses
http://www.life.uiuc.edu/micro/316/316resources/problems/recombination/mapping/crosses/

Unseen Life on Earth: A Telecourse
http://www.microbeworld.org/htm/mam/is_telecourse.htm

References

Allardet-Servent, A., S. Michaux-Charachon, E. Jumas-Bilak, L. Karayan, and M. Ramuz (1993). Presence of one linear and one circular chromosome in the *Agrobacterium tumefaciens* C58 genome. *J. Bacteriol* 175:7869-7874.

Bachmann, B. J. and K. B. Low (1980). Linkage map of *E. coli* K-12, edition 6. *Microbiological Reviews* 44:1-56.

Calley, J. N. and J. R. Jungck (2001). Microbial Genetics Construction Kit. In *The BioQUEST Library Volume VI*. Jungck, J. R. and V. G. Vaughan, Editors. Academic Press: San Diego, CA.

Curtiss, R. (1969). Bacterial conjugation. *Annual of Review Microbiology* 23: 69-136.

Hayes, William. (1968). *The Genetics of Bacteria and their Viruses*. John Wiley and Sons: New York.

Ibelgaufts, H. (1997). "Cats and Bugs." *Aliquotes V* (vii): (July).

Jacob, F., and E. L. Wollman (1961). *Sexuality and the Genetics of Bacteria*. Academic Press: New York.

Kalkum, M., R. Eisenbrandt, R. Lurz, E. Lanka (2002). Tying rings for sex. *Trends Microbiol* 10(8): 382-7.

Lederberg, J. and E. M. Lederberg (1952). Replica plating and indirect selection of bacterial mutants. *J. Bacteriology* 63: 399-406.

Lederberg, J. and E. L. Tatum (1946). Gene recombination in E. coli. *Nature* 158: 558.

Tatum, E. L. and J. Lederberg (1947). Gene recombination in the bacterium *Escherichia Coli*. *Journal of Bacteriology* 53(6):673-684.

Bibliography

Dale, J. W. (1998). *Molecular Genetics of Bacteria, Third Edition*. John Wiley & Son Ltd: New York.

Hahn, E., P. Wild, U. Hermanns, P. Sebbel, R. Glockshuber, M. Haner, N. Taschner, P. Burkhard, U. Aebi, S. A. Muller (2002). Exploring the 3D molecular architecture of *Escherichia coli* type 1 pili. *J Mol Biol* 323(5):845-57.

Maloy, S. R., J. E. Cronan, Jr., and D. Freifelder (1994). *Microbial Genetics, Second edition*. Jones and Bartlett Publishers: Boston, MA.

Figure and Table References

Figure 1.   Courtesy Joshua Tusin

Figure 2.   Courtesy Joshua Tusin

Figure 3.   Courtesy Joshua Tusin

Figure 4.   Courtesy Joshua Tusin

Figure 5.   From Bachmann and Low (1980)

Figure 6.   Screenshot from *Microbial Genetics Construction Kit*

Chapter 6

Activities for Video VI: Microbial Evolution

Microbial evolution poses some of the most challenging questions we face as humans. Research in this area increases our understanding of the distant origin of life on this planet and on the other hand, this same research suggests the control possibilities for surviving the rise of emergent, more resistant pathogens.

In this unit we can:

- compare highly conserved sequences in enolase protein variants,
- read a phylogenetic tree,
- explore a data set of West Nile Virus sequences from all over the world that date from the mid-20th century to the present, and
- compare sequence information for proteins produced by different organelles within the same organism.

Proteins: Historians of Life on Earth

What can the enzyme enolase reveal about the relationships between the organisms that produce this protein? Do variations in genetic sequence always alter both protein structure and protein function? Visualization tools and the Biology Workbench enable us to explore the active site on the enzyme.

Tree of Life: An Introduction to Microbial Phylogeny

How do phylogenetic trees help us look more closely at microbial evolution? Explore sequence data and extend your tree thinking to examine possible relationships between the microbes and other life forms. The Biology Workbench provides access to data and tools for investigating these relationships.

Tracking the West Nile Virus

How can viral sequences help us establish the origin of the virus that appeared in the US in 1999? Epidemiologists have adopted bioinformatics approaches using sequence data from strains of pathogens to track the movement of bacteria and viruses from continent to continent.

One Cell, Three Genomes

How can we investigate the phenomenon of endosymbiosis? Comparing nuclear and extra-chromosomal sequence data in photosynthetic eukaryotes provides data we can use to test hypotheses about the origins of mitochondria and chloroplasts.

Proteins: Historians of Life on Earth

Garry A. Duncan, Eric Martz, and Sam Donovan

Video VI: Microbial Evolution

Introduction

Prior to the 1980's, one of the most commonly accepted taxonomic hypotheses in biology was that all organisms belonged to one of two domains: (1) the eukaryotes, which included organisms whose cells contain a well-formed nucleus; and, (2) the prokaryotes, which included unicellular organisms whose cells lacked a nucleus, such as the bacteria. In recent years there has been a fundamental rethinking of how to organize the diversity of life. Recent molecular evidence has led to a new hypothesis—that the prokaryote domain is actually comprised of two distinct domains. Some bacteria-like organisms look like normal bacteria but may have had a distinct phylogenetic history. Consequently, these bacteria-like organisms may comprise a distinct domain, given the name Archaebacteria, or more simply, Archaea. The name reflects an untested conjecture about their evolutionary status. Recent phylogenetic evidence suggests that the Archaebacteria may be at least as old as the other major domains; hence, it now seems possible that the most recently categorized group of organisms may actually be the oldest. It is important to note that not all scientists agree with the three domain hypothesis. The bibliography section contains some suggested reading on this debate.

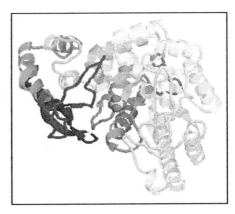

Figure 1. Yeast enolase (4ENL) structure. The helices and sheets are displayed as ribbons and the remainder of the molecule is displayed only as a backbone trace. The coloring shades from blue (amino terminus) to red (carboxyl terminus). There is a sulfate molecule, displayed as a CPK structure in the active pocket of the enzyme. (A color version is available on the *Microbes Count!* web site.)

Changes in the nucleotide sequence of DNA and amino acid sequence in proteins can be thought of as molecular fossils. That is, these changes act as historical records of evolutionary events and give us clues about the relatedness of different species in much the same way that changes in morphological characters, preserved in the form of fossils, give us clues about change over time. The extraordinary growth of sequence databases, along with the development of tools to explore and mine these databases, has radically enhanced the ability of biologists to uncover the patterns of organic evolution that have occurred throughout the history of life on Earth.

The following investigations can act as a springboard for you to pose evolutionary questions that might be answered by analyzing molecular data. To accomplish this end, it is important to have a user-friendly, web-based interface that enables you to access DNA and protein databases, perform alignments and construct phylogenetic trees. The *Biology Workbench* (hereafter known simply as the Workbench), developed at the National Center for Supercomputing Applications (NCSA), provides this user-friendly, web-based interface.

Investigation 1: Explorations in Evolution Through Protein Sequence Alignments and Phylogenetic Tree Construction

Objectives: (1) Gain experience using bioinformatics tools and databases, primarily through the *Biology Workbench*. (2) Use protein sequence data and analysis tools to evaluate the two hypotheses described above regarding different ways to classify the domains of life. In order to accomplish objective 2, a ubiquitous protein must be selected. For this investigation you will examine and compare the protein sequences of enolase, an enzyme involved in the last stage of glycolysis during which 3-phosphoglycerate is converted into pyruvate and a second molecule of ATP is formed. Enolase is found in all organisms, because they all utilize glycolysis to produce ATP for metabolism. You will compare the amino acid sequences of enolase from the seven species in Table 1 along with several species of your own choosing.

Table 1. Species information and web sites.

Species	General Information
Methanococcus jannaschii	Archaebacterium; thermophile (48-94 C); strict anaerobic that lives at pressures of over 200 atmospheres; autotroph that gets its energy from hydrogen and carbon dioxide producing methane. Source: http://jura.ebi.ac.uk:8765/ext-genequiz/genomes/mj/
Pyrococcus horikoshii	Archaebacterium; hyperthermophilic (optimal growth at 98 C, pH 7.0 and NaCl concentration 2.4%). Source: http://jura.ebi.ac.uk:8765/ext-genequiz/genomes/ph0004/
Escherichia coli	Bacterium; Gram negative; rod-shaped; facultative anaerobe; common inhabitant of the gut of warm blooded animals. Source: http://jura.ebi.ac.uk:8765/ext-genequiz//genomes/ec0005/index.html
Bacillus subtilis	Bacterium; Gram positive; rod-shaped; aerobe; nonpathogenic bacterium commonly found in the soil. Source: http://jura.ebi.ac.uk:8765/ext-genequiz//genomes/bs0005/index.html http://www.biojudiciary.org/glossary/index.asp?flt=b
Saccharomyces cerevisiae (yeast)	Unicellular eukaryote; fungus; economically important microbe because it has the ability to ferment glucose into ethanol and CO_2; its biochemistry and genetics are well known. Source: http://jura.ebi.ac.uk:8765/ext-genequiz//genomes/sc0006/index.html
Drosophila melanogaster (fruit fly)	Multicellular eukaryote; commonly known as the fruit fly; feeds on decaying plant matter. One of the best understood genetic organisms. Source: http://jura.ebi.ac.uk:8765/ext-genequiz//genomes/dm0006/index.html http://www.ceolas.org/fly/intro.html
Homo sapiens (human)	Multicellular eukaryote; mammalian primate; omnivore.

Overview of operations

Since you do not have an amino acid sequence of enolase for comparison, you must search for one. Once you have a sequence, you will do the following:

1. Generate a list of proteins with similar sequences by conducting a BLAST search for similar sequences;

2. Select a wide variety of species, representing all the major groups (Table 1, plus one or more selections of your own);

3. Create a multiple sequence alignment using your enolase sequences with ClustalW; and,

4. Construct a phylogenetic tree based on the sequence differences in the alignment. The *Biology Workbench* provides access to all of the databases and tools for these operations.

Using the *Biology Workbench*

The instructions below contain some of the information you will need to use the *Biology Workbench*. Please see the "Orientation to the *Biology Workbench*" document on the *Microbes Count!* web site for a broader overview of what the *Biology Workbench* is and how it is organized.

1. Entering the *Biology Workbench*:

   a. Launch your web browser and go to the following URL for the *Biology Workbench:* http://workbench.sdsc.edu/.

   b. If you have already set up an account on the Workbench, go to Step c now. If, however, this is your first time utilizing the Workbench, then click on the Click Here hyperlink to set up an account. Fill out the account information and click the Submit button and go to Step 2 below.

   c. Click on the hyperlink Enter the *Biology Workbench* 3.2

   d. Enter your user id and password and then click the Submit button.

   When working in the Biology Workbench, avoid using the browser's Back button; instead, use only the navigational buttons within the Workbench.

2. Starting a new session or resuming an old session:

   Before you can utilize the Workbench, you need to begin a New session or Resume a previous session, just as you need to begin a new file for word processing or continue a previous file in work processing. In other words, you cannot use the Protein Tools, Nucleic Tools, or Alignment Tools until you have resumed an old session or started a new session. Scroll down the page and click on the Session Tools button.

   a. To start a new session, click (i.e., select) Start New Session in the scrollable window and then click the Run button. On the new web page that appears, you need to name the session (= file) you are about to begin. In this case, we are going to name the session **Enolase** since we are going to be conducting a protein search, amino acid sequence alignment and tree construction for enolase. Now click the Run button. The page that now comes up is the same as the one that you were on a moment ago, except that your new session (i.e., Enolase, which is now a file name on a remote server) is now listed with your previous sessions, if you have any. (You may have to scroll down the page to see it.) If the radio button for the Enolase session is not already selected, click it now.

   b. You are now ready to begin searching for amino acid sequences. So, click the Protein Tools button near the top of the page.

3. Selecting a sequence:

   a. Now you are in the Protein Tools window. You need access to protein databases in order to perform your search. To do so, select Ndjinn – Multiple

Database Search in the scrollable window and then click the Run button. When the new web page appears, you need to indicate what protein you are searching for and what database(s) you wish to search. For this investigation, type enolase into the blank field following the word Contains. Now select the PDBFINDER database in the scrollable window, and click the Search button at the top of the web page. (You are selecting this database because the 3D structures are known for all of the proteins in this database.)

The Results page indicates that you have matched 15 unique records. (The number of unique records may be larger than 15 records since new records are being added on a continuous basis.) Click the box in front of the one that says: PDBFINDER:4ENL (from yeast). This will be the enolase sequence in which we will anchor the rest of our searches. (For later use in *Protein Explorer*, you will need to write down the PDB id number, which is the four digit alpha-numeric code—in this case, 4ENL.) Now click the Import Sequence(s) button, and continue to Step b.

b. Before going further, you should find out more about this enolase molecule. Click the box in front of the protein, select View Database Records of Imported Sequences from the scrollable window and then click the Run button. In the web page that now appears, select the Formatted radio button and then click the Show Record(s) button. You will find a wealth of information about this protein, including its amino acid sequence, its enzyme code number, citations, etc. You can even view the molecule in 3D (upper right of page). After viewing, click the Return button at the bottom of the web page.

4. Searching for records with similar sequences using *BLASTP* (Basic Local Alignment Search Tool for Proteins):

a. If it isn't already selected, click the box in front of pdbfinder:4enl_carbon-oxygen lyase.

b. In the scrollable window, select BLASTP – Compare a PS to a PS DB, and then click the Run button. Select the database SwissProt in the scrollable window. As you scroll to the bottom of the web page, you will note that you can specify a number of search criteria. For our purposes, we can use all of the default selections. At the bottom of the web page, click the Submit button. The BLASTP tool will find other similar protein sequences in the SwissProt protein database

5. Selecting records for alignment

a. Scroll down the BLASTP results page. For this activity, select the following six records (the yeast record, which you have already selected, acts as the seventh record and does not need to be selected again). In addition, select at least one or more species of your choice, and make a mental prediction (hypothesis) about where you think your species will fit on the phylogenetic tree.

Locus Name	Organism
ENO_DROME	*Drosophila melanogaster*
ENOB_HUMAN	Human
ENO_METJA	*Methanococcus jannaschii*
ENO_PYRHO	*Pyrococcus horikoshi*
ENO_BACSU	*Bacillus subtilis*
ENO_ECOLI	*E. coli*
Your sequences:	

b. Scroll back up to near the top of the web page and click the Import Sequences button. This action will import the amino acid sequences of all of the records (= sequences) you have selected. (The yeast record with which you started was already imported.) In the next step of the investigation, you will align the sequences.

6. Conducting an Alignment Using CLUSTALW – Multiple Sequence Alignment tool

a. Click the boxes of all of the records you wish to align, including the yeast record.

b. Select the CLUSTALW tool in the scrollable window. Now click the Run button. The CLUSTALW page appears, which contains all of the different settings you can alter when running this analysis. For this investigation, you will use all of the default settings. So, just click the Submit button. It will take the computer a few moments to develop the alignments.

c. Now scroll down the CLUSTALW Multiple Sequence Alignment page and see the alignments. At the top of the alignment, note the color coding key. The first alignment group contains the alignment for amino acids 1-60, while the second alignment group contains the alignment for amino acids 61-120, etc. You will also note some dashes within the alignment, which indicate missing amino acids. Scroll down below the alignment and you will note an unrooted tree. (We will reconstruct this tree in Step 7 below.) Continue scrolling and you will find additional information, including the number of amino acids in the enolases for each species. To save this alignment as a file for future viewing and for further analysis, scroll to near the bottom of the page, click the Import Alignments button.

7. Constructing a tree

a. In the Alignment Tools page, click the box in front of the CLUSTALW-Protein file of the aligned sequences. (You may have to scroll down the web page in order to see this box.) Selecting this box acts to select the entire list of records that have been aligned.

b. Select the *DRAWTREE* application tool in the scrollable window, and then click the Run button. (This tool draws an unrooted phylogenetic tree.) The *DRAWTREE* page appears, which contains all of the different settings you can alter. Again, we will use all of the default settings. Click the Submit button. Note that you can print out a copy of the tree, which should look similar to the one in Figure 2 below.

Figure 2. Unrooted tree. The branch points in the tree are called nodes, while the lines are called branches. The length of the branch is a direct measure of the amount of change that has occurred in this protein. Note that the two species of bacteria, *Bacillus subtilis* and *E. coli*, share a common branch of the tree; the two archaea, *Methanococcus jannaschii* and *Pyrococcus horikoshii*, share a common branch; and, the three eukarya—*Saccharomyces cerevisiae* (yeast), *Drosophila melanogaster*, and human—share a common branch.

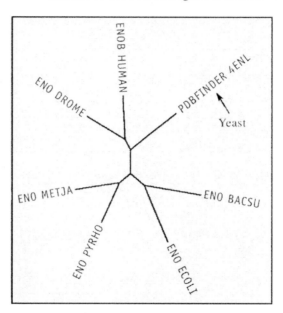

Questions for discussion:

1. Where would you expect *Methanococcus* and *Pyrococcus* to split off of the unrooted tree if the two domain (i.e., Bacteria and Eukarya) hypothesis is correct?

2. Where would you expect *Methanococcus* and *Pyrococcus* to split off of the unrooted tree if the three domain (i.e., Bacteria, Archaea and Eukarya) hypothesis is correct?

3. Did the species you added to the investigation appear on the tree where you predicted?

4. Which hypothesis does the tree in Figure 1 support?

8. Exporting sequence alignments in Fasta format

Click the Return button at the bottom of the web page in order to return to the Alignment Tools web page.

a. Click the box in front of the CLUSTALW-Protein file of the aligned sequences. In the scrollable window, select View Aligned Sequence(s) and then click the Run button. This will take you to the View window.

b. Scroll down the web page to view the sequences.

c. If you wish to import the sequence into *Protein Explorer*, you will need to change the format to Fasta. The Format window will probably say MSF; click the arrow for the dropdown window to open; select FASTA. This selection will automatically change the format to FASTA. Once the format has been changed, you can either Save the sequences to a file or Copy/Paste them into *Protein Explorer*. If you are going to continue on to Investigation 2 right now, then the Copy/Paste method is the easiest. Simply highlight and copy the entire group of sequences, including the > sign and the enzyme/species names, so that they can eventually be pasted into *Protein Explorer* (Step 3.c in Investigation 2 below).

Investigation 2: Visualizing the Evolution of Protein Structure in 3D

Objectives: (1) Become familiar with the many capabilities of *Protein Explorer* (PE); (2) be able to load a 3D structure of a protein into PE for viewing; (3) be able to place aligned sequences (Fasta format) into PE; and, (4) be able to visualize in 3D the evolutionary changes within the protein structure.

1. Opening *Protein Explorer (PE)*

Launch *Protein Explorer* in a new browser window at:

http://proteinexplorer.org.

On this web page, note all the different ways in which you can learn about PE, including a 1-2 hour tour that will give you a better idea of all of the capabilities of PE.

2. Loading the protein's 3D structure

a. Now go back to the PE home page (http://proteinexplorer.org). Locate the field where you enter the PDB Identification code. Type in the PDB ID for the enolase from yeast (i.e., 4ENL). (This is the code you were instructed to write down in step 3.a. in Investigation 1.) Now click the Go button. Be very patient. This takes awhile, particularly if this is the first time your computer has used PE. Click OK on any windows that come up. The 3D structure of the protein is now shown, but it does NOT indicate where there are any amino acid changes. This won't happen until the amino acid alignments you copied while working in the *Biology Workbench* (Step 8 in Investigation 1) are pasted into PE in step 3.c. below.

b. You should now be in the FirstView frame. Click the item Explore More with QuickView.

c. You should now be in the QuickViews frame. Near the bottom of this frame, click the hypertext where it says: Go to Advanced Explorer.

3. Pasting aligned sequence from *Biology Workbench* into *Protein Explorer*

a. You should now be in the Advanced Explorer frame. Click the item that says MSA3D: Multiple Sequence Alignment Coloring.

b. In the MSA3D Procedure frame, click item 5, which says Paste the alignment into the MSA3D ALIGNMENT FORM.

c. You are now in the MSA3D Alignment Form window. Place the cursor in the Alignment Box and Paste the alignment sequences (*Fasta* format) that you copied from the *Biology Workbench* (i.e., step 8 in Investigation 1 above).

d. Now copy/paste the yeast sequence (i.e., 4ENL), which is the sequence for which the 3D structure is known, from the Alignment Box into the 3D Sequence Box. The yeast sequence (i.e., 4ENL) will be in both the Alignment Box and the 3D Sequence Box.

e. Click the Color Alignment & Molecule button just below the 3D Sequence Box. In a moment, a new browser page will open, showing the color-coded alignments for all of the species. (This process may take several moments, so be patient.) The legend for the color codings is indicated at the top of the page. For example, green indicates that an amino acid at a specific position is identical for all species.

4. View the evolutionary changes of the protein in 3D:

a. If your screen is large enough, you will see the 3D structure rotating on another web page. Click on that web page to bring it to the front.

b. The backbone trace of enolase has been colored as indicated. The results are more easily appreciated when the full structure, including side chains, is shown with all atoms "spacefilled" (to van der Waals radii). In the MSA3D Result frame, click on each of the three links—Identical, Similar, and Different (i.e., the first three bullets)—to spacefill all categories. The red balls are water molecules. Click the Water button so that the red balls (water oxygens) are hidden, enabling you to clearly see the protein itself. The 3D model is showing the several billion year evolutionary history of enolase.

c. Point to the 3D model, click and hold down on the mouse button and move the mouse. This action allows you to rotate the molecule. The catalytic site is marked by a brown zinc (Zn) ion (nearly buried) and an easily spotted red-and-yellow sulfate ion that happens to be bound there. Note that the active site is entirely green (complete identity), showing billions of years of evolutionary conservation, while the peripheral region of the molecule is almost entirely yellow because of amino acid substitutions.

Questions for discussion:

1. What does the complete conservation of the amino acids in the active site suggest to you?

2. Why do you think the peripheral region of the enolase molecule has varied so much over time in contrast to the stability of the active site?

3. Are there other regions on the enolase molecule highly conserved, besides the active site? (Hint: are there conserved regions on the peripheral part of the molecule?) What might be the role of those regions?

4. Do other enzymes in glycolysis show similar results? (To answer this, you would have to repeat the investigation, substituting other glycolytic enzymes for enolase.

5. What other proteins might also be shared by the taxa used in the above investigations? You could investigate these proteins to determine whether or not they show the same pattern of evolution as enolase.

Protein Explorer's MSA3D is used in this activity because it helps you to understand the steps involved in coloring a protein molecule to show the rate at which each amino acid evolves. A much more automated, and therefore easier, method of identifying conserved and rapidly mutating residues for any 3D protein structure is the *ConSurf Server* (http://consurf.tau.ac.il), which also uses a more sophisticated and robust method for calculating conservation scores for each residue. It is less sensitive to the choice of sequences in the alignment than is MSA3D. Take a look at the ConSurf Gallery at its website if you are interested.

Web Resource Used in this Activity

Biology Workbench (http://workbench.sdsc.edu)

Originally developed by the Computational Biology Group at the National Center for Supercomputing Applications at the University of Illinois at Urbana-Champaign. Ongoing development of version 3.2 is occuring at the San Diego Supercomputer Center, at the University of California, San Diego. The development was and is directed by Professor Shankar Subramaniam.

Additional Resources

Available on the *Microbes Count!* web site at http://bioquest.org/microbescount

Text

A copy of this activity, formatted for printing

"Orientation to the *Biology Workbench*"

Related *Microbes Count!* Activities

Chapter 2: Searching for Amylase

Chapter 4: Molecular Forensics

Chapter 4: Exploring HIV Evolution: An Opportunity for Research

Chapter 6: Tree of Life: Introduction to Microbial Phylogeny

Chapter 6: Tracking the West Nile Virus

Chapter 6: One Cell, Three Genomes

Chapter 7: Visualizing Microbial Proteins

Unseen Life on Earth Telecourse

Coordinates with Video VI: Microbial Evolution

Relevant Textbook Keywords

Active site, Enolase, Glycolysis, Gram negative bacteria, Gram positive bacteria, Molecular evolution, Mutations, Phylogeny

Related Web Sites (accessed on 2/20/03)

Microbes Count! Website
http://bioquest.org/microbescount

Unseen Life on Earth: A Telecourse
http://www.microbeworld.org/htm/mam/is_telecourse.htm

Bibliography

Margulis, L. and K. Schwartz (1998). *Five Kingdoms: An Illustrated Guide to the Phyla of Life on Earth.* W. H. Freeman and Company, New York.

Hagen, J. (1996). Robert Whittaker and the Classification of Kingdoms. In *Doing Biology.* Harper Collins College Publishers. New York.

Figure and Table References

Figure 1.   Courtesy Sam Donovan

Figure 2.   Modified from *Biology WorkBench* (http://workbench.sdsc.edu)

Tree of Life: An Introduction to Microbial Phylogeny

Beverly Brown, Sam Fan, LeLeng To Isaacs, and Min-Ken Liao

Video VI: Microbial Evolution

Introduction

Bioinformatics tools allow microbiologists and evolutionary biologists to take a closer look at the evolutionary relationships between species by quantifying the differences between genetic sequences from many organisms. New hypotheses about the history of life on earth, including the re-rooting of existing phylogenetic trees–that is, identifying the earliest forms of life– are based on new sequence data that further defines the molecular similarities and differences between species

This activity will help familiarize you with the use of internet-accessible bioinformatics tools, methods, and data. You will have the opportunity to extend your understanding of phylogenetic relationships, explore the advantages and disadvantages of different molecular markers, and probe the complexity of establishing relationships between organisms.

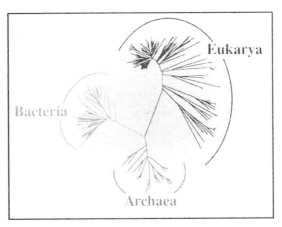

Figure 1. A phylogenetic tree showing the three domains, or superkingdoms, encompassing all life on earth.

Concepts

- The basics of constructing phylogenetic trees.

- The use of 16S rRNA and other important molecular markers for classification.

- Advantages and disadvantages of traditional vs. molecular methods for classification

Objectives

- Learn how to retrieve DNA sequence information using the internet.

- Understand how phylogenetic trees are constructed.

- Understand the kinds of information conveyed in a phylogenetic tree.

Background

Bioinformatics involves the use of computers to mine the vast amount of molecular information available to scientists. This information includes DNA and protein sequences for a wide variety of organisms. Phylogenetics, the development of hypotheses about the evolutionary history of a group of species, is an important application of bioinformatics.

Before the use of sequence data to develop evolutionary trees, phylogenies were based on morphological and metabolic characteristics of organisms. A phylogeny is typically represented as resolved or partially resolved bifurcating tree, consisting of nodes and branches (Figure 2a, 2b). The relative lengths of different branches

A *resolved* tree is one where every taxon has one and only one sister group. Under these conditions each branch *bifurcates*, or splits into two. A *taxon* is a group of organisms that is treated as a unit of analysis in a phylogenetic tree.

represent the genetic similarity , producing a weighted tree (Figure 2a). However, if the investigator is interested only in the respective groupings of species, sometimes the branches are shown as being of (arbitrary) equal length, yielding an unweighted tree (Figure 2b). If we know the root of an unweighted tree, we can draw it as a *cladogram*; like an unweighted tree, a cladogram's branch lengths are arbitrary. The numbers represent species which are nodes. There are also nodes where the lines join. The lines are branches.

Figure 2a. An unrooted weighted tree.

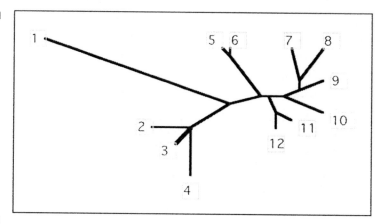

Figure 2b. An unrooted unweighted tree.

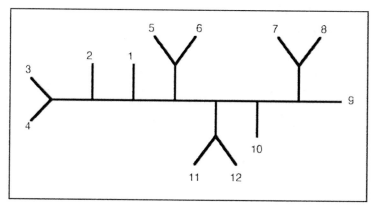

Evolutionary tree building seeks to find the tree that provides the most biologically sensible explanation of how diverse groups of species evolved. There can be many possible bifurcating tree shapes to consider for even a relatively small numbers of species. For example, the 8 taxa represented in Figure 3 below could be organized in 10,395 different unrooted trees.

Some methods of tree building use some secondary criteria to evaluate how well each of the possible tree shapes fit the data. However, an alternative approach that works better for larger numbers of species is to first construct a table containing measures of difference between pairwise comparisons of species (e.g. perhaps count the number of mutations that differ between the sequences of different species). The next step then uses the values in the table to determine the order in which species are clustered into a tree (Figure 3).

This procedure produces just one tree, and does not evaluate this tree against other possible trees, which may be almost as well supported by the data.

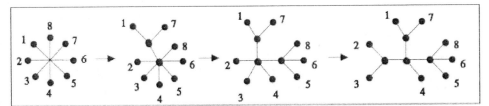

Figure 3. Phylogenetic tree building by pairwise comparisons. The starting assumption represented in the left-most tree is that all of the species are equally similar to one another. As different characters are considered, small groups of species are grouped, resulting in a tree that displays a hypothesis about how the species may be differentially related to one another (i.e., species 2 and 3 may be more closely related to each other than either of them is to species 1 or 7.)

There are several factors to consider when constructing a tree:

- Does this tree represent real relationships? Does it make sense?

- Can we expect the patterns in the data to produce a single unique tree?

- Can we infer a species phylogeny from the comparative analysis of only small sections of the genomes?

- Will protein or DNA or rRNA sequences be better for inferring phylogeny?

- Is it valid to compare homologous sequences if they are from different genome compartments (e.g. from chloroplasts, mitochondria, nuclei, nucleomorphs)?

- How far back in history can we reliably infer phylogeny from sequence data?

- What are the assumptions of the different tree-building methods? Which ones would be best? How would I decide?

Bioinformatics Tools

The basic resources consist of a bank of data, in our case RNA sequences, a program to retrieve sequences that we wish to use, and a program to compare the similarities and differences between sequences. These programs also display the comparisons of sequences in the form of a tree diagram. There are several online resources that provide access to these tools. For this activity we will use the *Biology Workbench*, developed at the National Center for Supercomputing Applications (NCSA). Note that simply generating the tree is not the end of the story. Interpretation is a time-consuming and extensive process, so you should expect to spend most of your time interpreting your trees.

The RNA sequences for this activity are stored in the GenBank database, an annotated collection of all publicly available DNA sequences. The GenBank records are organized into a variety of smaller databases, such as the GBBCT database that you will be searching. We will use the *Biology Workbench* to access the "Ndjinn" procedure. Using Ndjinn, we will retrieve sequences from GenBank databases by performing a text search for exact matches anywhere in the data record. We can search either by the sequence's GenBank accession number or by other characteristics such as species and gene names. In either case, searching with Ndjinn will return all records that contain the search text. Each record label in your search results will contain that record's accession number, so knowing the accession number of the particular sequence you're interested in can make it much easier to identify that sequence.

Beverly Brown, Samuel Fan, LeLeng To Isaacs, and Min-Ken Liao

Before the advent of molecular phylogenetics, relationships among organisms (and trees that summarize those relationships) were inferred from the organisms' morphological and metabolic characteristics. Now that the tools and data for analyzing DNA and protein sequences are available, it's also possible to examine the genetic relationships between microbes. In cases where molecular analysis yields trees similar to those obtained by non-molecular methods, we can gain added confidence that both data sets correctly describe the species' evolutionary history. In other cases, molecular and non-molecular trees may differ significantly, suggesting that processes such as convergent evolution may be occurring.

Overview of Operations

1. Choose the organisms that you will use to construct your phylogenetic tree.

2. Retrieve the sequences for your organisms.

3. Generate a phylogenetic tree from your database of sequences.

Selecting your organisms

The first step is choosing a diverse set of bacteria for your phylogenetic analysis. The 1984 edition of *Bergey's Manual of Systematic Bacteriology* (familiarly, *Bergey's*) is a four-volume reference that groups bacteria using primarily morphological and metabolic criteria (the 1984 edition is the last edition that does not extensively use molecular criteria). You can reasonably use the categories of species in each volume as the basis for building a morphological/metabolic tree.

Table 1 lists representative species from each major group in *Bergey's*, plus a eukaryotic example, *Saccharomyces cerevisiae*, for a total of eight groups. Choose at least two examples from each bacterial group, for a total of 14 species. You will also need to include *Saccharomyces cerevisiae*, the eukaryotic example, so you will end up with a total of 15 organisms for analysis. Enter the information for your organisms into the Species Data Table (Table 2). (The Species Data Table is also available on the *Microbes Count!* web site.) The GenBank numbers for the additional species in each group are not listed in Table 1 but you will learn how to find them in a later step.

Table 1. Representative species from each major group in Bergey's Manual of Systematic Bacteriology.

Volume 1A (Gram-negative bacteria)	
Escherichia coli	GBBCT #174375
Helicobacter pylori	GBBCT #402670
Salmonella typhi	GBBCT #2826789
Serratia marcescens	GBBCT #4582213
Treponema pallidum	GBBCT #176249
Additional species: *Agrobacterium tumefaciens, Bordetella pertussis, Thermus aquaticus, Yersinia pestis*	

Table 1. continued

Volume 1B (Rickettsias and endosymbionts)	
Bartonella bacilliformis	GBBCT #173825
Chlamydia trachomatis	GBBCT #2576240
Rickettsia rickettsii	GBBCT #538436
Additional species: *Coxiella burnetii, Thermoplasma acidophilum*	
Volume 2A (Gram-positive bacteria)	
Bacillus subtilis	GBBCT #8980302
Deinococcus radiodurans	GBBCT #145033
Staphylococcus aureus	GBBCT #576603
Additional species: *Bacillus anthracis, Clostridium botulinum, Lactobacillus acidophilus, Streptococcus pyogenes*	
Volume 2B (Mycobacteria and nocardia)	
Mycobacterium haemophilum	GBBCT #406086
Mycobacterium tuberculosis	GBBCT #3929878
Additional species: *Mycobacterium bovis, Nocardia orientalis*	
Volume 3A (Phototrophs, chemolithotrophs, sheathed bacteria, gliding bacteria)	
Anabaena sp.	GBBCT #39010
Cytophaga latercula	GBBCT #174236
Nitrobacter winogradskyi	GBBCT #402722
Additional species: *Heliothrix oregonensis, Myxococcus fulvus, Thiobacillus ferrooxidans*	
Volume 3B (Archeobacteria)	
Methanococcus jannaschii	GBBCT #175446
Thermotoga subterranea	GBBCT #915213
Additional species: *Desulfurococcus mucosus, Halobacterium salinarium, Pyrococcus woesei*	
Volume 4 (Actinomycetes)	
Actinomyces bowdenii	GBBCT #6456800
Actinomyces neuii	GBBCT #433527
Actinomyces turicensis	GBBCT #642970
Eukaryotic representative	
Saccharomyces cerevisiae	GBPLN #172403

Table 2. The Species Data Table

Volume 1A (Gram-negative bacteria)	
	GBBCT #
	GBBCT #
Volume 1B (Rickettsias and endosymbionts)	
	GBBCT #
	GBBCT #
Volume 2A (Gram-positive bacteria)	
	GBBCT #
	GBBCT #
Volume 2B (Mycobacteria and nocardia)	
	GBBCT #
	GBBCT #
Volume 3A (Phototrophs, chemolithotrophs, sheathed bacteria, gliding bacteria)	
	GBBCT #
	GBBCT #
Volume 3B (Archeobacteria)	
	GBBCT #
	GBBCT #
Volume 4 (Actinomycetes)	
	GBBCT #
	GBBCT #
Eukaryotic representative	
Saccharomyces cerevisiae	GBPLN #172403

- Assuming that Bergey's classification accurately reflects microbial evolutionary relationships, sketch an unrooted tree for the species in your Species Data Table. You will need to make and justify judgments about which major groups are most closely related and which are more distant.

Online access to sequence databases and alignment and tree-building software.

The instructions below contain the basic information that you will need to use the *Biology Workbench*. Please see the "Orientation to the *Biology Workbench*" on the *Microbes Count!* web site for a broader overview of what the

Biology Workbench is and how it is organized. You may also want to look at the "Proteins: Historians of Life on Earth" activity in Chapter 6 of this book.

Entering the *Biology Workbench*

Go to <http://workbench.sdsc.edu>. If you do not already have a *Biology Workbench* account, click on the Click Here hyperlink to set up an account. Fill out the account information and click the Submit button. Once you have established an account, click on the link Enter the *Biology Workbench*, enter your user ID and password, and begin a new session.

> *When working in the Biology Workbench, avoid using the browser's Back button; instead, use only the navigational buttons within the Workbench.*

If you leave the *Biology Workbench*, use the following directions to continue your unfinished work: Log into the *Workbench*. Click on "Session Tools." Then highlight "Resume Session" in the scroll box. Click "Run." The name of your folder should appear near the top of your screen.

Building a sequence database

1. You are now ready to begin retrieving the sequences for your organisms. Click on the Nucleic Tools button near the top of the page.

2. Highlight Ndjinn – Multiple Database Search and click Run. This will bring up a page that gives you access to the many databases available through *Biology Workbench*. To do a search you will need to indicate what you are searching for and which database you wish to search.

   In this project, we will use sequence data for the small protein subunit of ribosomal RNA (rRNA) from the species listed in Table 1. Ribosomal RNAs are labeled based on their sedimentation rates (S), which relates to their size. The small subunit in prokaryotes involves a 16S rRNA while eukaryotes have a slightly larger 18S rRNA. In retrieving sequence data for yeast, therefore, we will search for the 18S gene within the GBPLN database (GenBank Plant Sequences, which includes the fungi and algae). Sequence data for the 14 bacterial species, by contrast, will involve searching for the 16S gene within the GBBCT database (GenBank Bacterial Sequences).

3. Start by retrieving the gene sequence for the yeast.

   - Type in "*Saccharomyces cerevisiae* AND 18S". This is the sole representative of Eukaryotes.

   - Select the GBPLN database. (You have to scroll down for this option.)

   - Click on Show 10 hits. Drag down and select Show 50 hits.

   - Click on Search. You will see a list of choices. Scroll down until you see gbpln:172403. Highlight this line and click on the Import Sequence(s) button located at the end of the first line of the interactive box. This will import the gene sequence for this microorganism.

   Alternatively, since you already know the GenBank number for the sequence you want, you can instead base your search on this number rather than on the species and gene names. Both this approach and the one described above will turn up multiple records containing the search text; changing the search option to Begins with rather than Contains can help reduce the number of spurious hits.

4. Now you will follow the same procedure to retrieve the sequences for the rest of your microorganisms from GenBank. Repeat the following steps for each of the bacterial species that you entered into your Species Data Table:

    •   Highlight Ndjinn and click Run.

    •   Since these are bacterial species, select the GBBCT (GenBank Bacterial Sequences) database.

    •   Type in the name of your organism but this time use the 16S rather than the 18S gene.

    •   Click on Search. Scroll down until you see the entry for your bacteria. (If the organisms you chose from Table 1 did not have GenBank numbers, this is where you will find their GenBank numbers. Enter the number into your Species Data Table now.) Highlight the entry and click on the Import Sequence(s) button.

At the end of your searches, you should have 14 bacterial sequences and one yeast sequence.

Constructing a tree

Now you are ready to generate an unrooted tree.

5. Conducting an alignment using ClustalW – Multiple Sequence Alignment tool

    •   Highlight Select All Sequences in the scroll box. Click Run. All the boxes in front of the organism names should be checked.

    •   Scroll down and highlight CLUSTALW – Multiple Sequence Alignment. Click Run. The ClustalW page will appear. There are a number of different settings on this page; for this analysis you can simply use all of the default settings.

    •   Click Submit. The screen will go blank and you may have to wait several minutes. Wait until a screen titled "CLUSTALW" with "Sequence alignment" appears.

Scroll down to examine the DNA sequences and how they align with each other. Make sure that the ends of the bacterial sequences do not have gaping holes (more than 20 or 30 bases). If a few do not align well, representing incomplete sequences or sequences beyond the rRNA coding region, you should delete them from the Ndjinn field and go back and import other sequences until you find some that align better. Otherwise your comparison will be invalid.

6. Scroll further down and you will see your tree! It should be similar to the tree in Figure 3.

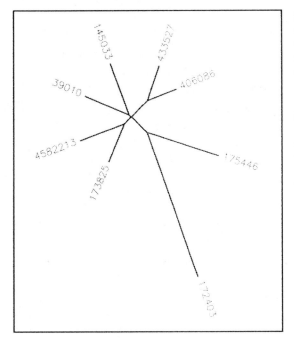

Figure 3. A weighted unrooted tree including a single species from each group in Table 1. Note that the longest branch is associated with the eukaryotic outgroup (S. cerevisiae, GB#172403), implying that of all sequences present, this sequence has the lowest genetic similarity to the others. Species labels correspond to GenBank accession numbers; in analyzing your own tree, you may wish to add species names(see text).

Saving your tree for further analysis

7. You will need to copy your tree into a word processing application so that you can study it. The exact procedure for doing this will vary depending on the type of computer you are using and on your browser. In general, click on the tree image and hold until a menu appears. Highlight Copy this Image and release.

8. Now open Microsoft *Word* and open a document. You may want to choose the document that already has your Species Data Table in it. Choose Paste from the Edit menu and your tree will appear. You may want to adjust the size of your tree by selecting the tree image on one of the lower corners and dragging. Be sure to give your tree a label (such as Figure 1) and include a short description.

9. Your tree is labeled using the GenBank numbers for each species. It will be easier for you to interpret your tree if you add the species name. Look up each GenBank number in your Species Data Table and write the corresponding species name beside the number. You can do so by choosing TextBox under the Insert menu. Click where you want the textbox to be located and type the name of the species. Drag the textbox next to the corresponding number in your tree.

Exercise 1: Traditional classification with the *Bergey's* manual

- Discuss the similarities and differences between the tree you sketched earlier in the activity and the computer generated trees. How do the morphological/metabolic relationships in the former relate to the molecular relationships identified in the latter?

Exercise 2: Molecular classification

One way to test the robustness of a phylogenetic tree is to draw a new tree after adding or subtracting one or more species from the analysis. This procedure may drastically change trees that aren't well-supported by the data, while leaving better-supported trees relatively unaltered. In the next two exercises, you will employ this technique to evaluate your tree.

A. Based upon the molecular tree you've generated previously, predict what the tree might look like if you added two more bacterial species of your choice from Table 1. Where do you expect these new species to appear on the tree?

B. Add the two bacterial species you chose in part A. Use the *Biology Workbench* to generate a tree. (Save a copy of the tree to your Word document as Exercise 2.) Does this agree with what you expected? Does it make sense? Why or why not?

C. Which of the morphological and metabolic groups from *Bergey's* seem to be consistent with the molecular data? Which aren't? Use colors to indicate the organisms within each major group.

Exercise 3:

A. Based upon your tree from Exercise 2, predict what the tree might look like if you removed one of the microorganisms listed in your Species Data Table.

B. Remove the microorganism you chose in part A. Use the *Biology Workbench* to generate a tree. (Save a copy of the tree to your Word document as Exercise 3.) Does this agree with what you expected? Does it make sense? Why or why not?

C. Did any microorganisms change their positions? Do any of them still stay together? Explain.

Critical thinking questions

On Taxonomy

- *E. coli* and *S. typhi* are closely related based on their positions in the tree, which indicates that they have similar genomic makeup. Does this mean they also have similar metabolic characteristics?

- *B. subtilis* is a spore former. How does spore formation benefit a microbe? How can you classify spore formers morphologically? What if the spore is an endospore?

- *S. marcescens* produces red pigments which are not expressed at all temperatures. Is this a reliable characteristic for identification under all conditions? How could pigment production be used as a reliable characteristic?

- Both *Mycoplasma* and *Haemophilus* can cause respiratory diseases. Does this piece of information help scientists determine their phylogenetic relationship?

- *E. coli* and *E. coli* 0157:H7 are the same species. Why do they behave so differently? What kinds of genetic differences might be responsible? How could you test these hypotheses?

On Phylogenetics

- Why do we use 18S rRNA information for yeast and 16S for all the other prokaryotes?

- Why do we need to classify organisms? Why are phylogenetic trees important?

- What are important characteristics for classification? How would you choose? How are you going to get the information on the characteristics?

- Can you predict the metabolic characteristics of your unknown based on its position on the tree?

- You are the scientist on the expedition to the recently discovered planet of Lebesamin. Considering that Lebesamin is over 6 parsecs from Terra and given the atmospheric differences between the two planets, there is remarkable similarity in the genetic makeup of the life forms on the two planets. You are responsible for collecting and identifying all microorganisms on the planet. Using the skills you have developed through this exercise, write up a proposal for classifying the organisms.

Web Resource Used in this Activity

Biology Workbench (http://workbench.sdsc.edu)
  Originally developed by the Computational Biology Group at the National Center for Supercomputing Applications at the University of Illinois at Urbana-Champaign. Ongoing development of version 3.2 is occuring at the San Diego Supercomputer Center, at the University of California, San Diego. The development was and is directed by Professor Shankar Subramaniam.

Additional Resources

Available on the *Microbes Count!* web site at http://bioquest.org/microbescount

Text

A copy of this activity, formatted for printing

"Species Data Table" in MSWord format

"Orientation to the *Biology Workbench*"

Related *Microbes Count!* Activities

Chapter 2: Searching for Amylase

Chapter 4: Molecular Forensics

Chapter 4: Exploring HIV Evolution: An Opportunity for Research

Chapter 6: Proteins: Historians of Life on Earth

Chapter 6: Tracking the West Nile Virus

Chapter 6: One Cell, Three Genomes

Chapter 7: Visualizing Microbial Proteins

Unseen Life on Earth Telecourse

Coordinates with Video VI: Microbial Evolution

Relevant Textbook Keywords

Archaea, Bacteria, Eukarya, Evolution, Nucleic acids, Phylogenetic relationships, rRNA

Related Web Sites (accessed on 2/20/03)

Microbes Count! Website
http://bioquest.org/microbescount

Unseen Life on Earth: A Telecourse
http://www.microbeworld.org/htm/mam/is_telecourse.htm

References

Holt, J. G., Editor-in-Chief (1984) *Bergey's Manual of Systematic Bacteriology, Volume 1-4.* 1st Edition. Williams & Wilkins:Baltimore. http://www.cme.msu.edu/bergeys/pubinfo.html

Figure and Table References

Figure 1. Courtesy Mitchell Sogin, Josephine Bay Paul Center, Marine Biological Laboratory

Figure 3. Modified from *Biology WorkBench* (http://workbench.sdsc.edu)

Tracking the West Nile Virus

Erica Suchmann and Mark Gallo

Background

West Nile virus was first isolated in 1937 from a woman in Uganda; however, scientists did not characterize the life cycle of the virus until 1951 during an outbreak in Egypt. West Nile virus, a member of the family Flaviridae, is carried by *Culex* sp. mosquitoes. These mosquitoes prefer to feed on birds, but also bite humans, horses and other mammals. In 1957, doctors identified West Nile virus as a cause of meningoencephalitis (swelling of the brain and spinal column) in humans and, in 1960, as a cause of disease in horses.

Until recently West Nile virus had never been reported in the Americas. That all changed in 1999 when illnesses due to West Nile virus infection in humans and birds were identified in New York City. Microbiologists and epidemiologists initially hoped that the New York outbreak would be limited to one summer and that the virus and the mosquitoes would be unable to overwinter in the cold New York climate. Their hopes were dashed when human cases were identified in 12 states in 2000. By the end of 2002, 3984 human cases had been identified in 39 states, and avian, animal and mosquito cases had been identified in 44 states (Centers for Disease Control web site: http://www.cdc.gov/ncidod/dvbid/westnile/; January, 2003)

Most people bitten by West Nile virus-infected mosquitoes do not develop symptoms. For the 20% that do, the symptoms of West Nile virus infection include a mild fever, headache, and body chills that begin between 3-14 days after initial exposure. In most people these symptoms last for a few days before resolving with no lasting effects. Less than 1% of those infected with West Nile virus go on to develop encephalitis or meningitis, but these severe symptoms seem to be more likely to develop in people over 50.

To determine the origin of the virus isolated in the New York City outbreak, scientists at the Centers for Disease Control (CDC) constructed a database of genetic sequences of the West Nile virus from strains that have been isolated during outbreaks since 1951. Discovering how the virus travels and enters new environments greatly enhances the ability of researchers to predict where the virus may spread in future outbreaks. The same techniques for studying the movement of the virus through populations can also be used to study outbreaks of other diseases and to trace biological agents used for terrorist attacks.

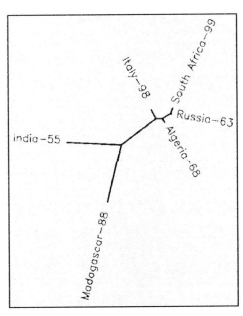

Figure 1. An unrooted tree based on aligned sequences for the envelope glycoprotein gene in six different strains of the West Nile virus. Each strain is labeled by the country and year it was collected. The tree was generated by running a ClustalW using the *Biology WorkBench*.

Epidemiological Activity

Your task is to look at genetic sequences from West Nile virus cases in mammals and birds and to compare them in order to determine the likely origin of the strain found first in the United States. If we assume that mutations occur continuously as the virus replicates in various hosts, then having access to the genetic sequences from several West Nile virus strains found in different locations will be helpful. We can compare these sequences and use their relatedness to hypothesize about how West Nile virus arrived in New York.

Figure 2. An example of the nucleic acid sequence data for the envelope glycoprotein gene in the West Nile virus (strain India-55).

```
>India-55
CCGACTACAGTCGAATCGCATGGCATCTACTCAACACAGCAGGGCGCCACTCAAGCAGGCCGGTTCAGCA
TAACTCCAGCCGCGCCATCATACACTCTAAAGCTTGGGGAATATGGAGAAGTCACTGTGGACTGCGAGCC
CCGCTCGGGCATAGACACAAACGCGTACTATGTGATGACAGTTGGGACGAAGACGTTTCTAGTCCACCGA
GAATGGTTCATGGACCT
```

The CDC's West Nile virus sequences are available to the public in the GenBank Viral Sequences (GBVRL) database. This database can be accessed from the *Biology Workbench* at the University of California San Diego Supercomputing Center (http://workbench.sdsc.edu). The bioinformatics tools you will need to look for patterns and relationships will be available there as well.

For an overview of the *Biology Workbench* and how it is organized, please see the "Orientation to the *Biology Workbench*" document on the *Microbes Count!* web site. You may also want to take a look at the "Proteins: Historians of Life on Earth" and "Tree of Life: An Introduction to Microbial Phylogeny" activities in Chapter 6 for some examples of using sequence data to create a phylogenetic tree.

Once you're set up, create a Session called West Nile and then upload the CDC sequences in the file called "West Nile Sequences.doc", located in the West Nile folder on the *Microbes Count!* web site. Align the sequences using ClustalW. An unrooted phylogenetic tree will be produced. The shape of the tree, the length of the branches, and the position of the different strains of West Nile provide information about relatedness; e.g., strains within a cluster are closely related to each other.

Figure 3. The ClustalW tool aligns the sequences, enabling you to compare West Nile virus collected from various geographical areas. Only partial sequences for the envelope gene in a subset of the CDC strains are shown here.

```
South_Africa-99_  CCAACCACTGTGGATTCGCATGGTAACTACCCCACACAGATTGGGGCCACTCAGGCAGGG
Russia-63_        CCAACCACTGTGGAGTCGCATGGAAACTACCCCACACAGATTGGGGCCACTCAGGCAGGG
Algeria-68_       CCAACCACTGTGGAGTCGCATGGAAACTACCCCACACAGATTGGGGCCACTCAGGCAGGG
Italy-98_         CCAACCACTGTGGAGTCGCATGGAAACTACTCCACACAGATTGGGGCCACTCAGGCAGGG
India-55          CCGACTACAGTCGAATCGCATGGCATCTACTCAACACAGCAGGGCGCCACTCAAGCAGGC
Madagascar-88_    CCGACGACTGTTGAATCTCATGGCAATTATTCAACACAGGTTGGGGCCACCCAGGCTGGA
                  ** ** ** ** ** ** ***** *  * *   * ****** ** ***** ** ** **
```

1. What are your conclusions about the origin and path of the West Nile virus that found its way to New York in 1999? Defend your methods as well as your answers.

2. Compare your tree to the tree published in the paper "Origin of West Nile Virus responsible for an Outbreak of Encephalitis in the NE US" (Lanciotti, 1999).

   Why might your tree differ from the one published in the paper?

3. Design an experiment to determine the geographic spread of West Nile virus in the USA? What data and kinds of information would you need? What factors (e.g. political borders, population density, climate, etc.) do you expect to have the largest impact? Explain.

4. The tree only shows the relationship between strains of the virus for the envelope gene. What are some reasons for why the scientists at the CDC chose the envelope glycoprotein sequences to compare?

Web Resource Used in this Activity

Biology Workbench (http://workbench.sdsc.edu)

Originally developed by the Computational Biology Group at the National Center for Supercomputing Applications at the University of Illinois at Urbana-Champaign. Ongoing development of version 3.2 is occuring at the San Diego Supercomputer Center, at the University of California, San Diego. The development was and is directed by Professor Shankar Subramaniam.

Additional Resources

Available on the *Microbes Count!* web site at http://bioquest.org/microbescount

Text

A copy of this activity, formatted for printing

"West Nile Sequences.doc", a text file for use with the activity

"Orientation to the *Biology Workbench*"

Related *Microbes Count!* Activities

Chapter 2: Searching for Amylase

Chapter 4: Molecular Forensics

Chapter 4: Exploring HIV Evolution

Chapter 6: Proteins: Historians of Life on Earth

Chapter 6: Tree of Life: Introduction to Microbial Phylogeny

Chapter 6: One Cell, Three Genomes

Chapter 7: Visualizing Microbial Proteins

Unseen Life on Earth Telecourse

Coordinates with Video VI: Microbial Evolution

Relevant Textbook Keywords

Epidemiology, Pathogen, Virus

Related Web Sites (accessed on 3/20/03)

American Society for Microbiology
http://asmusa.org

Centers for Disease Control, West Nile virus information
http://www.cdc.gov/ncidod/dvbid/westnile/

Microbes Count! Website
http://bioquest.org/microbescount

Unseen Life on Earth: A Telecourse
http://www.microbeworld.org/htm/mam/is_telecourse.htm

References

Lanciotti, R. S., J. T. Roehrig, V. Deubel, J. Smith, M. Parker, K. Steele, B. Crise, K. E. Volpe, M. B. Crabtree, J. H. Scherret, R. A. Hall, J. S. MacKenzie, C. B. Cropp, B. Panigrahy, E. Ostlund, B. Schmitt, M. Malkinson, C. Banet, J. Weissman, N. Komar, H. M. Savage, W. Stone, T. McNamara, and D. J. Gubler (1999). Origin of the West Nile virus responsible for an outbreak of encephalitis in the northeastern United States. *Science* 286 (5448):2333-2337.

Bibliography

Campbell G. L., A. M. Marfin, R. S. Lanciotti, and D. G. Gubler (2002). West Nile virus. *Lancet Infectious Diseases* 2:519-29.
http://www.cdc.gov/ncidod/dvbid/westnile/resources/wnv-campbell-etal.pdf

Jai X-Y, T. Briese, I. Jordan, A. Rambaut, et al. (1999). Genetic analysis of West Nile New York 1999 virus. *Lancet* 354:1971-1972.

Scherret, J. H., M. Poidinger, J. S. Mackenzie, A. K. Broom, V. Deubel, W. I. Lipkin, T. Briese, E. A. Gould, and R. A. Hall (2001). The relationship between West Nile and Kunjin viruses. *Emerging Infectious Diseases* 7(4).
http://www.cdc.gov/ncidod/eid/vol7no4/scherret.htm

Figure and Table References

Figure 1. Modified from *Biology Workbench* (http://workbench.sdsc.edu)

Figure 2. Modified from *Biology Workbench* (http://workbench.sdsc.edu)

Figure 3. Modified from *Biology Workbench* (http://workbench.sdsc.edu)

One Cell, Three Genomes

John R. Jungck, Sam Donovan, and John M. Greenler

One of the major hypotheses in biology in the past thirty-five years is the "endosymbiotic theory" which asserts that eukaryotic intracellular organelles originated from bacterial ancestors (Figure 1). Even though the concept of endosymbiosis has been around since the 1880's, it was not until the 1960's when Lynn Margulis popularized the notion that mitochondrial and chloroplast genes are more closely related to genes in specific currently living bacteria than they are to nuclear genes that the concept was systematically tested. Margaret Dayhoff and her colleagues at the National Biomedical Research Foundation (Schwartz and Dayhoff, 1978) initially tested this theory by building phylogenetic trees from protein sequences and found that such trees were consistent with the endosymbiotic hypothesis.

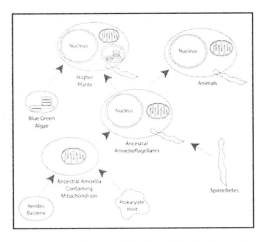

Figure 1: Endosymbiotic theory for the bacterial origin of eukaryotic organelles.

Phylogenetic trees built from newly available protein and nucleic acid sequence data both support and confound our understanding of endosymbiosis. One paradox is that all known chloroplast-encoded proteins are part of larger multimer complexes, of which at least one other subunit is nuclear-coded. How can chloroplasts have originated in prokaryotic cells if their functional genes are located in nuclei not just in the chloroplasts themselves?

In this activity, you will develop and test hypotheses about the endosymbiotic theory and both horizontal gene transfer and protein trafficking between chloroplasts, mitochondria, nuclei, and prokaryotic cells. Using the tools of bioinformatics you will be able to compare multiple sequence alignments and phylogenetic trees for different but related proteins to investigate possible evolutionary pathways and impacts of natural selection on interacting proteins.

Background

Numerous protein and gene trees have been generated to test the endosymbiotic theory. Analyses, made by drawing upon the rapidly expanding, massive, publicly available databases of protein and nucleic acid sequences, have repeatedly demonstrated that mitochondria share close ancestry with the purple bacteria such as *Rhodosprilium rubrum* and chloroplasts share a majority of their gene sequences with the photosynthetic blue green algae such as *Prochloron* (Figure 2). Sugita et al. (1997) illustrated the robustness of this conclusion by combining data on 22 ribosomal protein genes in one phylogenetic analysis.

Figure 2. A phylogenetic tree of life with mitochondria, chloroplast, nuclear genomes and their common ancestry.

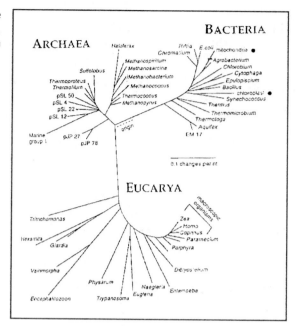

A typical eukaryotic green alga or land plant cell has genomes from three different ancestries: nuclei, mitochondria and chloroplasts. The DNA of each of these three organelles can be studied independently and, thus, provide additional information on the evolutionary history of a species. Because the genomes of mitochondria and chloroplasts are much smaller than nuclear genomes, they are much easier to sequence completely and, thus, many more of their genomes have been published. Barkman et al. (2000) have asserted that one of the advantages of studying all three genomes is that: "Plant phylogenetic estimates are most likely to be reliable when congruent evidence is obtained independently from the mitochondrial, plastid, and nuclear genomes with all methods of analysis." Both Chaw et al. (2000) and Wang et al. (2000) have also effectively employed this approach to estimate times of divergence in pine species as well as to construct their phylogenies.

An early argument against the endosymbiotic theory was based upon the observations that chloroplasts code for only a few proteins and that most of the proteins found in these organelles are actually encoded for by the nuclear DNA. A second argument emphasized that chloroplasts with such depauperate genomes have not been shown to be able to live on their own when isolated from the eukaryotic cell. Despite these objections, the endosymbiotic theory is overwhelmingly accepted by the majority of biologists. One surprising inference was that while some bacterial genes had been lost in the process of evolution, in most cases these genes had been transferred to nuclei of the respective host organism. This process is known as xenologous gene transfer. Furthermore, frequently a gene that encoded one subunit of an oligomeric protein remained in the organelle while another gene that encoded a different subunit(s) of that same oligomeric protein was transferred to the nucleus.

Genes and proteins move around in the cell at different time scales (Figure 3). The xenologous transfer of genes between organelles within the same cell occurs over the span of many generations and can be inferred from phylogenetic analysis of sequences from multiple, distantly related species. On the other hand, protein trafficking occurs during the life cycle of a single cell. Typically cell biologists have extensively studied protein trafficking, but not the gene trafficking which occurs within a much longer time frame.

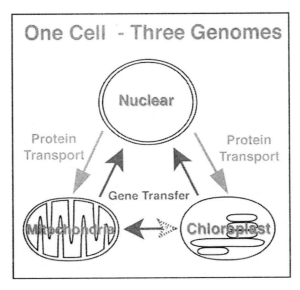

Figure 3 Protein and gene trafficking in a plant cell.

Hartl and Clark (1997) have summarized the rates at which both structural and nucleotide changes occur in different organelles in different taxa (Table 1). Surprisingly, chloroplasts and mitochondria not only have different evolutionary origins, but they also continue to diverge at qualitatively different rates.

Genome	Nucleotide Substitution Rate	Structural Evolution Rate
Flowering plant chloroplast DNA	Slow	Slow
Flowering plant mitochondrial DNA	Slow	Fast
Mammalian mitochondrial DNA	Fast	Slow
Fungal mitochondrial DNA	Fast	Fast

Table 1: Comparative rates of change in chloroplast and mitochondrial genomes over long periods of evolution (From Hartl and Clark (1997).

Problem

In this activity, we would like you to explore the "Chloroplast Paradox" (Table 2). What are some of the preliminary considerations you need to think about before you begin to investigate the products of gene trafficking? If two subunits of a protein work cooperatively as a whole, even though one is coded in a different

Table 2: A Chloroplast Genome Paradox

A Chloroplast Genome Paradox

*   All known chloroplast encoded proteins are a part of larger multimer complexes, of which at least one other subunit is nuclear encoded. Intracellular protein trafficking brings the subunits into a common space before they function at high capacity.

*   The chloroplast genome is much smaller than that of bluegreen bacteria; most chloroplast genomes only contain around 120 genes.

*   The function of at least 40 of the proteins corresponding to this chloroplast genome has yet to be determined.

organelle from where it functions, shouldn't the evolutionary pressures on the two subunits be parallel and compensatory? If so, how could we examine this possibility? Unlike the diversity in structural rates illustrated in Table 1, presumably we should observe a similar evolutionary divergence in a phylogenetic reconstruction of the history of the two subunits. Thus, we should expect the topology of the trees constructed for each subunit to be both congruent with one another (that is, we should be able to rearrange them geometrically without changing their topology to be superimposable on top of one another).

In the previous activities in Chapter 6, you have been introduced to the *Biology WorkBench* and the construction of phylogenetic trees. Here we would like you to simultaneously develop multiple sequence alignments for two different proteins that are found in a number of species. You can then determine whether the phylogenetic history of the species is congruent with that which we might infer from these two gene trees. The technique for examining two topologies at the same time was developed by Eisenberg and is called "phylogenetic profiling" (Figure 4).

Figure 4: Phylogenetic Profiling. (In the figure, the blue protein is on the upper left, the green protein is on the upper right, and the red protein is below.)

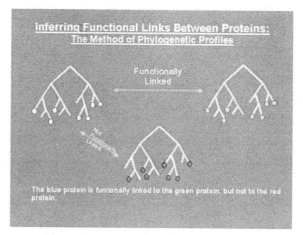

Activities

I. We will begin by choosing a protein to study and a set of photosynthetic algae to study it in. As you could infer from Table 2, there are about 80 chloroplast proteins that have both a nuclear counterpart and a cyanobacterial ancestor which you could study. Here we will begin with one of the most famous ones, namely Rubisco (ribulose bis phosphate carboxylase/oxygenase). See Figure 5. This enzyme is associated with the important step within photosynthesis where exogenous carbon is assimilated into an organic molecule. The carbon fixation reaction can be found in your textbook (also see Figure 6).

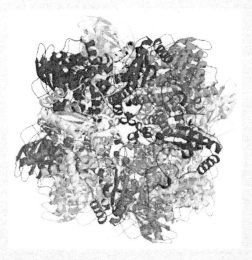

Figure 5: Ribbon diagram of activated ribulose-1,5-bisphosphate carboxylase (Rubisco) from the green alga *Chlamydomonas Reinhardt II* complexed with 2-carboxyarabinitol-1,5-bisphosphate.

Based upon the 1.84Å structure 1IR2 (http://www.rcsb.org).

Because Rubisco has been so extensively studied, it is very well represented in protein sequence databases and has been the subject of numerous phylogenetic studies. Its three dimensional structure has been worked out for several species. However, as far as we know, to this date no one has published a phylogenetic profiling analysis of Rubisco. The genes that code for Rubisco are located in both the chloroplast and nucleus of photosynthetic organisms.

- Using Biology Workbench, develop phylogenetic trees for both nuclear and chloroplast genes that determine Rubisco. Make a phylogenetic profiling analysis of the trees constructed from the two subunits.

You can choose a species to work with from the collection of RubiscoL and RubiscoS sequences that are provided on the Microbes Count! web site. Your work will be easier if you remember the following points:

- First, clean up your sequences so that you are only using multiple sequence alignments for a fairly constant length (except for the variation caused by internal insertions and deletions).

- Second, you should not try to build trees that are too deep because xenologous transfer has not only moved the subunit genes, but early horizontal gene transfers between species destroys monophyletic trees and turns them into networks that are more difficult to interpret.

The sequences that we have placed on the web site have been cleaned up and selected to avoid too many difficulties in analysis.

II. Compare sequences for Rubisco in photosynthetic bacteria, single-celled photosynthetic eukaryotes, and plants. What can you learn about evolutionary changes from these comparisons? How does your choice of organisms affect the trees you derive?

Extend your analysis by comparing each set of subunits with corresponding sequences in cyanobacteria. Do you support the hypothesis that chloroplasts evolved only once in plants or is your analysis consistent with the idea that there have been serial endosymbiotic events?

III. Choose a mitochondrial gene that codes for a protein involved in cellular respiration and develop a phylogenetic tree for it as well, using as many of the same species from the previous trees as possible. In this case we have not provided you with sequences so please take similar precautions to those described above.

How do the phylogenetic trees for genes that encode parts of the same protein compare? If you find differences, how can you explain these differences? How do your explanations fit with the endosymbiotic theory? Can you develop explanations based upon the endosymbiotic theory that explain your findings as well?

How does the phylogenetic tree based on mitochondrial DNA compare to the previous trees? What explanations can you suggest for similarities and differences you observe? Evolutionary biologists test these hypotheses by looking at sequence data and protein structures. Go to the references to see how others have approached these problems as you probe you questions further.

IV. A radical interpretation of the endosymbiotic theory is that animals, fungi, and plants are bacteria that live as colonies within a common cell membrane as well as in numerous collections of these superorganismal organizations. Rubisco is the most abundant protein on earth with a wide phylogenetic distribution (every green plant and alga). Another problem associated with Rubisco that you might investigate is photorespiration because it reduces the overall energetic efficiency of photosynthesis (Figure 6). This phenomenon occurs because diatomic oxygen is able to bind in the active site of Rubisco instead of carbon dioxide; you could think of diatomic oxygen as a competitive inhibitor of the reaction and, hence, the reaction is very much influenced by the concentration of diatomic oxygen in the air.

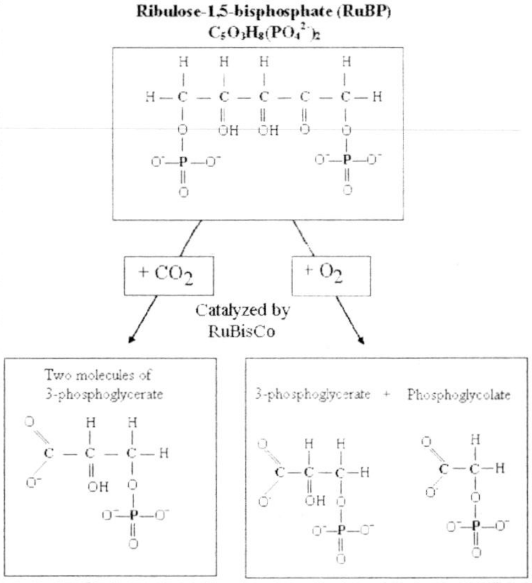

Figure 6: A comparison of two dark reactions catalyzed by Rubisco: fixation of carbon dioxide in photosynthesis (left: 5C + 1C = 2(3C)) and photorespiration (right: 5C = 3C + 2C).

Ecologically, in temperate climatic habitats, early wet, cool spring plants (C3) are more susceptible to the effects of photorespiration than arid, hot late summer plants (C4). Thus, the natural selection pressures on the selective affinity of various substrates (diatomic oxygen and carbon dioxide) for Rubisco is an interesting structural and evolutionary problem that has agricultural and forestry ramifications. Several researchers have sought to be able to double the production of biomass worldwide by altering the gene for this one enzyme in many species of crop plants.

V. Additional challenges:

- Choose other chloroplast and mitochondrial enzymes that have subunits encoded by genes in the nuclei of single-celled photosynthetic eukaryotes and in land plants. Do you observe differences when you compare phylogenies of photosynthetic algae with those of plants?

- What is the impact of including non-photosynthetic, parasitic plants, which may have relaxed selection pressures on Rubisco, in your data set?

These problems can become the basis for extended study of the evolution of organelles in eukaryotic organisms using bioinformatics tools.

Web Resources Used in this Activity

Biology WorkBench

The *Biology WorkBench* was originally developed by the Computational Biology Group at the National Center for Supercomputing Applications at the University of Illinois at Urbana-Champaign. Ongoing development of version 3.2 is occurring at the San Diego Supercomputer Center, at the University of California, San Diego. The development was and is directed by Professor Shankar Subramaniam.

Platform Compatibility: Requires an internet connection and a current web browser.

Additional Resources

Available on the *Microbes Count!* web site at http://bioquest.org/microbescount

Text

A copy of this activity, formatted for printing

"Orientation to the *Biology Workbench*"

Data

RubiscoL and RubiscoS sequences from nuclei and chloroplasts. Full Rubisco from cyanobacteria.

Related *Microbes Count!* Activities

Chapter 2:   Searching for Amylase

Chapter 4:   Molecular Forensics

Chapter 4:   Exploring HIV Evolution: An Opportunity for Research

Chapter 6:   Proteins: Historians of Life on Earth

Chapter 6:   Tree of Life: Introduction to Microbial Phylogeny

Chapter 6:   Tracking the West Nile Virus

Chapter 7:   Visualizing Microbial Proteins

Relevant Textbook Keywords

Carbon fixation, Chloroplast, Endosymbiotic theory, Eukaryote, Genome, Horizontal gene transfer, Mitochondrion, Nucleus, Oligomer, Organelle Photosynthesis, Phylogenetic profiling, Phylogenetic tree (phylogeny), Protein trafficking, Protomer, Rubisco (ribulose bis phosphate carboxylase/oxygenase), Xenologous

Unseen Life on Earth Telecourse

Coordinates with Video VI: Microbial Evolution

Related Web Sites (accessed on 3/20/03)

Biology WorkBench
http://workbench.sdsc.edu

Microbes Count! Website
http://bioquest.org/microbescount

Unseen Life on Earth: A Telecourse
http://www.microbeworld.org/htm/mam/is_telecourse.htm

References

Barkman T. J., G. Chenery, J. R. McNeal, J. Lyons-Weiler, W. J. Ellisens, G. Moore G, A. D. Wolfe, and C. W. dePamphilis (2000). Independent and combined analyses of sequences from all three genomic compartments converge on the root of flowering plant phylogeny. *Proc. Natl. Acad. Sci. U. S. A.* 97(24):13166–13171.

Chaw, S. M., C. L. Parkinson, Y. Cheng, T. M. Vincent, and J. D. Palmer (2000). Seed plant phylogeny inferred from all three plant genomes: monophyly of extant gymnosperms and origin of Gnetales from conifers. *Proc. Natl. Acad. Sci. U. S. A.* 97(8):4086–91.

Hartl, D. L. and A. G. Clark (1997) *Principles of Population Genetics.* Sinauer:Sunderland, MA.

Lederberg, J. (1997). Infectious disease as an evolutionary paradigm. *Emerging Infectious Diseases* 3(4):October-December.

Marcotte E. M., I. Xenarios, A. M. van Der Bliek, and D. Eisenberg (2000). Localizing proteins in the cell from their phylogenetic profiles. *Proc. Natl. Acad. Sci. U. S. A.* 97(22):12115–12120.

Pellegrini, M., E. M. Marcotte, N. J. Thompson, D. Eisenberg, and T. O. Yeates (1999). Assigning protein functions by comparative genome analysis: protein phylogenetic profiles. *Proc. Natl. Acad. Sci. U. S. A.* 96(8):4285–4288.

Qiu, Y-L, J. Lee, F. Bernasconi-Quadroni, D. E. Soltis, P. S. Soltis, M. Zanis, E. A. Zimmer, Z. Chenz, V. Savolainen, and M. W. Chase (1999). The earliest angiosperms: evidence from mitochondrial, plastid and nuclear genomes. *Nature* 402:404-407.

Schwartz, R. M. and M. Dayhoff (1978) Origins of prokaryotes, eukaryotes, mitochrondria, and chloroplasts. *Science* 199:395-403.

Sugita, M., H. Sugishita, T. Fujishiro, M. Tsuboi, C. Sugita, T. Endo, and M. Sugiura (1997). Organization of a large gene cluster encoding ribosomal proteins in the cyanobacterium *Synechococcus* sp. strain PCC 6301: comparison of gene clusters among cyanobacteria, eubacteria and chloroplast genomes. *Gene* 195(1):73–79.

Wang, X. Q., D. C. Tan, and T. Sang (2000). Phylogeny and divergence times in *Pinaceae*: evidence from three genomes. *Molecular Biology and Evolution*. 17(5):773–781.

Bibliography

Delwiche, C. F. and J. D. Palmer (1996). Rampant horizontal transfer and duplication of rubisco genes in eubacteria and plastids. *Molecular Biology and Evolution*. 13:873-882

Kellogg, E. A. and N. D. Juliano (1997). Structure and function of RuBisCO and their implications for phylogenetic studies. *American Journal of Botany* 84:413-428.

Williams, S. E., V. A. Albert and M. W. Chase (1994). Relations of *Droseraceae*: a cladistic znalysis of rbcL sequence and morphological data. *American Journal of Botany*. 81.

Figure and Table References

Figure 1.   Courtesy Joanna Cramer and John M. Greenler

Figure 2.   Modified from Lederberg, J. (1997)

Figure 3.   John M. Greenler

Figure 4.   Courtesy Todd O. Yeates (University of California-Los Angelos)

Figure 5.   Courtesy Stephen J. Everse (University of Vermont)

Figure 6.   Courtesy Tia Johnson

Table 1.   Modified from Hartl and Clark (1997)

Chapter 7

Activities for Video VII: Microbial Diversity

The diversity of the microorganisms we encounter compels us to consider new ways of looking at these ubiquitous life forms. Microbes occupy more of the earth's divergent living spaces than other life forms. Ironically, the microbial populations we cannot see transform most of what we can see. To study the differences between microbes, we examine populations through their metabolic products, growth patterns, and even genetic sequence information.

In this unit we can:

- model the effects of reducing nutrient loads to surface waters on microbial populations and oxygen levels within the Mississippi River basin and Gulf of Mexico Hypoxia Zone,
- mathematically analyze the leaves of the leopard plant whose "spots" result from in situ viral populations, and
- investigate enzymes produced by diverse microbes using 3D structural models and sequence data.

Microbiology of Stratified Waters

Sampling of aquatic microorganisms reveals a surprising feature of many open water systems - stratification. Planktonic and benthic microbes occupy distinct zones in stratified water columns. Explore how the benthic consumer populations respond to environmental cues such as changes in sunlight, wind effects, or increases in organics due to pollution with the model HypoxiaZone.

Valuing Variable Variegation

Microbes would remain under-explored if we limited our observations to those that could be successfully cultured in the laboratory. Computational methods are useful with field populations as well as with colonies in a Petri dish. To determine if a pattern of a viral infection in a leaf is random, we will examine data provided by image capture and visualization techniques.

Visualizing Microbial Proteins

What kinds of information do we expect to find in a structural model of a protein? Do sequence data provide similar information? A comparison of a CGTase and an alpha-amylase using both of these approaches provides evidence of functional similarities between the two enzymes. Search extensive online databases and explore additional proteins with bioinformatics tools.

Microbiology of Stratified Waters

Ethel D. Stanley, Howard T. Odum, and Elisabeth C. Odum

Video VII: Microbial Diversity

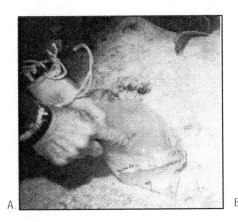

A.

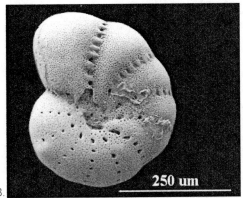

B.

Figure 1. A. Collecting live foraminifera. These organisms are useful as indicators of local environmental conditions. B. Opportunistic foraminifera such as *Cribroelphidium poeyanum* d'Orbigny tolerate low oxygen environments and anthropogenic pollution. (Images courtesy of Pamela Hallock Muller, College of Marine Science. University of South Florida.)

Scientific investigations in the Gulf of Mexico have documented a large area of the Louisiana continental shelf with seasonally-depleted oxygen levels (< 2mg/l). Most aquatic species cannot survive at such low oxygen levels. The oxygen depletion, referred to as hypoxia, begins in late spring, reaches a maximum in midsummer, and disappears in the fall. After the Mississippi River flood of 1993, the spatial extent of this zone more than doubled in size, to over 18,000 km² and has remained about that size each year through midsummer 1997. The hypoxic zone forms in the middle of the most important commercial and recreational fisheries in the coterminous United States and could threaten the economy of this region of the Gulf.

Hypoxia in the Gulf of Mexico (2001) National Centers for Coastal Ocean Science Gulf of Mexico Hypoxia Assessment

In deeper lakes and estuaries and even in some open seas, waters become stratified with a layer of warm water on top blocking oxygen from reaching the denser, colder or more saline water below. When upper waters are nutrient rich, excess organic matter sinks down and is decomposed by bacteria in the deeper waters. The decomposition uses up oxygen, which cannot be replaced due to the stratification, and the deeper waters become anaerobic. Temperature affects the rate at which decomposition and oxygen use occurs. In the winter, when the winds are stronger and surface waters cool off, the stirring that results may mix the layers. Later the waters stratify again.

Although hypoxia zones are natural phenomena, increasing pollution from human activities and natural events is affecting the size and persistence of these zones in ways that threaten to disrupt the natural ecology of the systems. Scientists are currently investigating a large area of anaerobic water in the Gulf of Mexico. In

Figure 2. The *HypoxiaZone* model of stratified waters in an ocean, estuary or lake ecosystem with upper and deep waters.

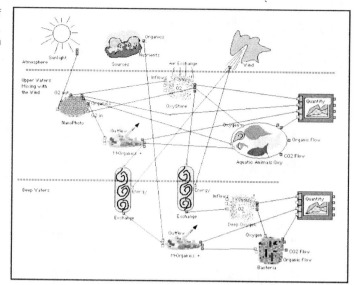

addition to the organic matter produced by photosynthesis in the upper layer, sediments coming out of the Mississippi-Atchafalaya River Basin sink into the deeper water. Oxygen levels drop as surplus organic matter is broken down by bacteria and other benthic consumers such as the foraminifera in Figure 1.

We can use the model *HypoxiaZone* (shown in Figure 2) to investigate the producer and consumer populations in terms of their metabolic exchanges in organics and oxygen levels in stratified waters. The elements in *HypoxiaZone* are represented by picture icons and the flows between the elements are shown by connecting lines. (See Table 1.)

One important limitation in the model is that nutrients are held constant in the upper level so producers achieve carrying capacity in response to available sunlight. The flow of organics in sediments does impact consumers in the deeper waters.

You can investigate *HypoxiaZone* further by changing the variables in the model. You can increase or decrease the wind, add a flow of organic matter from outside the system, increase the turbidity of the water (by reducing the sunlight), or vary the starting quantities of several other variables. The default values of the *HypoxiaZone* model variables are included in Table 1.

Table 1. Elements of the "Hypoxia Zone" model.

Block	Description	Default Value
Sunlight	**Sunlight** Represents a constant sunlight intensity as set in the dialog box.	The value is set at 4000 kilocalories per square meter per day.
Organics Nutrients Sources	**Sources** Represents the flow of organics from city and farmland.	Organics are set at 0. Inorganics are set at 1.

Ethel D. Stanley, Howard T. Odum, and Elisabeth C. Odum

Table 1. Continued

Block	Description	Default Value
Wind	**Wind** Represents a constant flow that mixes waters using the physical energy from wind or other sources.	Stirring energy is set at 0.1.
O2 out / Organics / O2 in / Phytoplankton	**Phytoplankton** Represents the microbial producers	Starting biomass is set at 1.
Air Exchange / Inflows / O2 / OxyStore	**Air Exchange** Represents the storage of dissolved oxygen in water. The amount of oxygen exchanged across the water surface depends on the concentration of oxygen in the water.	Initial oxygen storage is 8%. Air exchange rate set at 1.
Outflow / Organics +	**Organics** Represents stored organics. In this model the organics come from the photosynthesis of the plankton in the upper waters and the Sources block. Organics flow out in proportion to the quantity stored.	Starting storage is 5. Upper ouflow rate is set at 5. Lower outflow rate is set at 1.
Oxygen / Organic Flow / CO2 Flow / Aquatic Animals Oxy	**Aquatic Animals** Represents the fish and other animals that use the organics for food and fuels. They also use oxygen (02) for their respiration. CO2 is produced by their respiration and released into the air.	Starting quantity is set at 1
Exchange	**Exchange** Represents quantities of organic matter or dissolved oxygen exchanged within the water. The exchange is in proportion to the difference in concentration	None. Stirring is caused by the wind value.
Inflows / O2 / Deep Oxygen	**Deep Oxygen** Represents the storage of dissolved oxygen isolated in waters. There is no exchange with the air or other flows.	Initial storage set at 5%.
Oxygen / CO2 Flow / Organic Flow / Bacteria	**Bacteria** Represents microbial consumers of which bacteria are the most abundant.	Starting quantity set at 1.

To start the model, open the file "HypoxiaZone.mox", located on the *Microbes Count!* web site. Run the simulation using the *Extend* software. (More detailed instructions for opening, running, saving, and modifying models can be found in "Getting Started with Extend" on the *Microbes Count!* web site.)

Figure 3. The simulation graph of atmosphere and upper waters on the left shows initial increases in photosynthesis and organics with corresponding decrease in oxygen. Fish (Aquatic Animals) remain at a predetermined level to simplify the model. The simulation graph of deep waters on the right shows initial decreases in oxygen, organics, and microbial populations. No photosynthesis occurs at this depth.

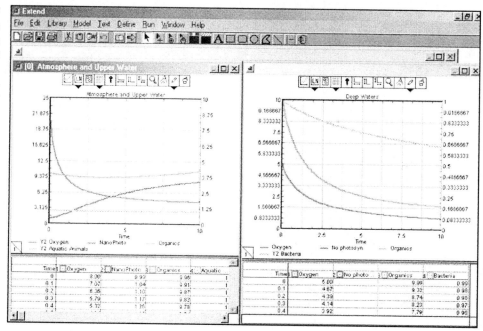

Carefully study the results of your initial run of the *HypoxiaZone* simulation. It will look like Figure 3. You will want to save and/or print your results for comparison with other runs. Use the Notebook included in the Extend program or simply capture screenshots and paste them in a standard word processing file. (See "Getting Started with Extend" for more details on how to do this.)

Now consider the following questions about the *HypoxiaZone* model:

1. What happens to the microbial populations of both producers and consumers in this simulation?

   What changes do you see for organics in both upper and deep waters?

   If you run the simulation for more cycles, what happens to the populations of upper water producers and deeper water consumers? (Hint: Select Run from the Menubar and hold the mouse button down to select Simulation Setup. Change the End simulation time to 100 and click on Run Now.)

2. Explore the effects of wind on the water system. (The End simulation time should still be set to 100. Then double click on the Wind icon to changes the value to .01). What happens to the oxygen in the deep waters? How can you account for this change?

   What happens if there is no wind (0)?

3. You can model the impact of an unusual geophysical event such as a volcanic eruption that results in persistent cloudiness. (Hint: Double click the Sun icon to enter 3000 for the sunlight intensity.) Describe what happens to the microbial populations in both the upper and deeper waters.

   You can model the effect of other seasonal sunlight fluctuations or changes in turbidity by increasing or decreasing the intensity. Create your own scenario and model it. Explain how your new model differs from the default model.

   What did you learn about the populations, oxygen, or organics use?

4. Investigate the effect of flooding on a stratified water system. Specifically, look at the impact on deeper water microbial populations when a large quantity of organic sediments is washed from the river directly into the deep organic matter. (Hint: Double click the Sources icon and set organics flow to 1.) Describe your results.

Extended Activities

1. Monitoring water systems includes routine sampling of planktonic and benthic populations. Provide a rationale for the use of foraminifera as indicator species for assessing environmental stresses like hypoxia. What can a survey of forams tell you about past fluctuations in the water systems? Cite your sources.

2. Make a case for the reduction of nutrient loads to surface waters within the Mississippi River basin and Gulf of Mexico. (Hint: Data on the Gulf of Mexico hypoxia zone is readily accessible from published scientific data and articles on the internet.) How can you use the *HypoxiaZone* model to support this position?

3. Models are used routinely to help us understand aspects of a complex system, but models are necessarily simpler representations of the system. Are there aspects of the problem above that you think are critical, but can not model?

   Describe one and explain why it can not be modeled with *HypoxiaZone*.

Software Used in this Activity

Hypoxic Zone

> Howard T. Odum (University of Florida) and Elisabeth C. Odum (Santa Fe Community College)

> Platform Compatibility: Macintosh and Windows

Additional Resources

Available on the *Microbes Count!* web site at http://bioquest.org/microbescount

Software

Hypoxia Zone

Text

> A PDF copy of this activity, formatted for printing

> "Getting Started with Extend"

Related *Microbes Count!* Activities

> Chapter 3: Modeling Wine Fermentation

> Chapter 3: Biosphere2: Unexpected Interactions

> Chapter 10: Making Sense of the Complex Life Cycles of Toxic Pfiesteria

Unseen Life on Earth Telecourse

> Coordinates with Video VII: Microbial Diversity

Relevant Textbook Keywords

> Anaerobe, Hypoxia zone, Indicators, Organic matter

Related Web Sites (accessed on 4/13/03)

Microbes Count! Website
> http://bioquest.org/microbescount

Microbial Indicators
> www.lu-ces.org/documents/SOKreports/ microbialindicators.pdf

Unseen Life on Earth: A Telecourse
> http://www.microbeworld.org/htm/mam/is_telecourse.htm

References

EPA Addresses Hypoxia in the Gulf
> www.epa.gov/msbasin/hypoxia.htm

Foraminifera as bioindicators in coral reef assessment
> http://www.marine.usf.edu/reefslab/foram/html_files/titlepage.htm

Hypoxia in the Gulf of Mexico (2001). National Centers for Coastal Ocean
    Science Gulf of Mexico Hypoxia Assessment
    http://www.nos.noaa.gov/products/pubs_hypox.html

Odum, H. T. and E. C. Odum. (2002) *Hypoxia Zone.* Contact the BioQUEST
    Curriculum Consortium (bioquest@beloit.edu) for information.

Figure and Table References

Figure 1.   Courtesy of Pamela Hallock Muller, College of Marine Science,
             University of South Florida

Figure 2.   Screen shot from *HypoxiaZone* model

Figure 3.   Screen shot from *HypoxiaZone* model

Valuing Variable Variegation

John R. Jungck, Brooke Halgren, Joshua Tusin, and Tia Johnson

Video VII: Microbial Diversity

Using Natural Experiments to Understand Viral Infection

Scientists are often faced with the challenge of studying complex processes in natural environments. While the tools of experimental biology rely on the manipulation of variables, many important phenomena occur in situations where manipulation is difficult if not impossible.

Figure 1. The leopard plant *Ligularia tussilaginea 'Aureomaculata.'*

What is an experiment? The word experiment conjures up notions of manipulated variables with controls. But how do scientists test hypotheses outside of the controlled laboratory? How can we take advantage of natural experiments to test hypotheses about complex problems? What can we learn from astrophysicists, archaeologists, and evolutionary biologists about observing natural systems in systematic ways to test hypotheses in natural experiments?

Too often we move biology into the lab in order to simplify and control the variables, yet we want to return to the outdoors to apply what we have learned. With contemporary technology, it is easy to go directly into the field to test hypotheses in natural settings.

Consider how we could mathematically analyze microbes and viruses that grow without our intervention and manipulation. Historically microbiologists have characterized populations of bacteria by inoculating known media and counting the resulting colonies on Petri plates. Similarly, a standard mathematical analysis of viruses is carried out by counting plaques. But how do we characterize *in situ* infection, such as the virus responsible for the variegated leaves of the leopard plant *Ligularia tussilaginea* in Figure 1? (Color versions of all photographs are available in the PDF of this activity, located in the Variable Variegation folder on the *Microbes Count!* web site.)

Horticulturists and amateur gardeners have long valued variegation – so much so that whole books are devoted to them. A popular variegated plant introduced from Japan during the Victorian era, the leopard plant, was extensively used in garden borders and in greenhouses for its bold speckled appearance. The white and yellow blotches are actually caused by a viral infection. Usually we consider an infestation to be something to be eradicated–a sign that the plant is sick or less attractive. Yet horticulturists have carefully nurtured and propagated a number of variegated varieties with good economic success. Virally infected tulips that exhibit color breaks are highly valued for their patterns of pigment effects, as illustrated in Figure 2 on the next page. Similarly infected bulbs were bought and sold for fortunes during "tulipomania" in Europe in the 1670's (Pollan, 2001).

Figure 2: Two varieties of tulips. The one on the right is highly valued for its variegated appearance, which is due to a virus. The virus can be lost if winters are too severe. This may have been what was responsible for a massive drop in the value of tulips on the world markets at the end of the 1670's.

Natural Experiments

We can consider the pattern of spotting on the leaves of the leopard plant, *Ligularia*, as a natural experiment. Examining the viral populations found on the leopard plant in Figure 1, we see the leaves vary considerably from all green to heavily covered by infection spots. How can we make sense of this rich variegation?

Figure 3. This image of a *Ligularia* leaf was photographed with fluorescent cold light. The light spots are caused by viral infection.

In this activity you will use a leaf of the leopard plant, or another similarly variegated plant, to explore the premise that the pattern of infection is random. As you observe the distribution of viral infection on the *Ligularia* leaves shown in Figure 3 and on the additional samples available on the *Microbes Count!* web site, consider the following questions:

- What does random mean in this context?

- What characteristics would you consider in this exploration?

- Could the investigation of characteristics such as size, color intensity, hue, distribution, smoothness of borders, distances between blotches or from veins, position with respect to the stomata on the leaf, or developmental age of the leaf affect our assumptions?

- What tests could we perform?

The BioQUEST Curriculum Consortium

John R. Jungck, Brooke Halgren, Joshua Tusin, and Tia Johnson

- What data collection methodology should we consider? Are the data parametric (continuous numbers) or non-parametric (can we simply rank order by qualitative descriptors)?

Devise and carry out an experiment to test the relationship between viral infection and characteristics of the plant host. The following two examples may suggest avenues for exploration:

Researcher Illo Hein, in 1926, reported that: "The rounded yellow spots of the leaf of *Ligularia Kaempferi* Sieb. & Zucc. (*Farfugium grande* Lindl.) show a gradual diminution in the amount of green color from the solid green cells at the border to the almost colorless central tissue. The area consists visibly of more or less definite concentric zones, which range from normal green through various tones of green to yellow, and in the larger spots to white. This pattern is very suggestive of Liesegang rings...." Looking at the leaves under more powerful magnification allows us to look for rings as noted by Hein, and to analyze the distribution of stomata within sites of infection.

More recently, research in Brazil and in the United States has focused on understanding the natural history of infection with *Xylella fastidiosa*, a bacterium that infects orange trees. In infected plants *Xylella fastidiosa* physically occludes the xylem and hence interrupts the transport of fluids and nutrients. The disease it causes is called citrus variegated chlorosis (CVC). CVC produces a very distinctive pattern of variegation along all the veins. The genome of *Xylella fastidiosa* was sequenced in Brazil in 2000, and progress is being made toward developing methods of pathogen control (Hagmann 2000). Could our understanding of the *Ligularia* case help in developing methods to limit infection and crop loss?

For your experiment you may use the images of *Ligularia* on the web site, or use plants from your local nursery. A rough estimate of the area of yellow blotches and/or leaves can be done by using an overlay of transparent graph paper. However, you can increase both the efficiency and accuracy of the estimations by using freeware tools of quantitative image analysis, such as *NIH Image*, developed by Wayne Rasband and available online. The *NIH Image* equivalent for PC users, *Scion Image*, is available for free download from Scion Corporation. For more information on area measurement of multiple blotches at one time, consult the "Laboratory Procedures" document on the *Microbes Count* web site. this document will guide you in using *NIH Image* to analyze the leaves in your "natural experiment."

When you design your experiment, keep in mind the following questions:

- What specific features of pathogenesis and/or viral reproduction might influence observed results?

- What is your hypothesis?

- What analytical tests will you perform?

John R. Jungck, Brooke Halgren, Joshua Tusin, and Tia Johnson

Acknowledgements:

The authors wish to thank Sue Risseeuw, Deborah Sapp-Lynch, Analyne Schroeder, Amanda Sanders, and Joanna Cramer for their assistance in the preparation of this module.

Software Used in this Activity

NIH Image

Developed by Wayne Rasband at the U.S. National Institutes of Health and available for download at: http://rsb.info.nih.gov/nih-image/

Platform Compatibility: Macintosh only

Scion Image

Developed by Scion Corporation and available for download at: http://www.scioncorp.com

Platform Compatibility: Windows only

Additional Resources

Available on the *Microbes Count!* web site at http://bioquest.org/microbescount

Text and Data

A PDF copy of this activity, formatted for printing

"Laboratory Procedures" document

Digital photographic images of leaves from three *Ligularia tussilaginea* "Aureomaculata"

Related *Microbes Count!* Activities

Chapter 1. Modeling More Mold

Chapter 2. Shaped to Survive

Chapter 9. Citrus Canker: Alternatives for Control

Unseen Life on Earth Telecourse

Coordinates with Video VII: Microbial Diversity

Relevant Textbook Keywords

Ligularia tussilaginea, Leopard Plants, Plant viruses, Variegation, *Xylella fastidiosa*

Related Web Sites (accessed on 4/14/03)

History House story on Tulipomania
http://www.historyhouse.com/in_history/tulip/

Microbes Count! Website
http://bioquest.org/microbescount

Origin, Development, and Propagation of Chimeras
   http://aggie-horticulture.tamu.edu/tisscult/chimeras/chimeralec/chimeras.html

Terra Nova Nurseries for information on plant cloning, tissue culture, and
   variegated varieties
   http://terranovanurseries.com

Unseen Life on Earth: A Telecourse
   http://www.microbeworld.org/htm/mam/is_telecourse.htm

References

Hagmann, M. (2000). Intimate portraits of bacterial nemeses. *Science* 288: 801.

Hartvigsen, G. (2000). The analysis of leaf shape using fractal geometry.
   American Biology Teacher 62 (9): 664-668.

Hein, I. (1926). Changes in plastids in variegated plants. *Bulletin of the Torrey
   Club* 53: 411-419.

Kistler, R. A. (1995). Image acquisition, processing & analysis in the biology
   laboratory. *American Biology Teacher* 57 (3): 151-157.

Pollan, M. (2001). *The Botany of Desire*. Random House: New York.

Bibliography

Blystone, R. and R. A. Cooper (2001). Image analysis using starfish embryos.
   In *The BioQUEST Library Volume VI*. Jungck, J. R. and V. G. Vaughan,
   Editors. Academic Press: San Diego, CA.

Zaniello, T. (1988). *Hopkins in the Age of Darwin*. University of Iowa Press:
   Iowa City.

Figure and Table References

Figure 1.   Tia Johnson

Figure 2.   Joshua Tusin and Tia Johnson

Figure 3.   Tia Johnson

Visualizing Microbial Proteins

Ethel D. Stanley and Keith D. Stanley

Video VII: Microbial Diversity

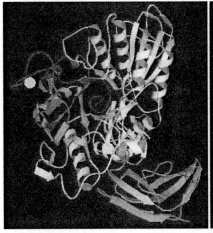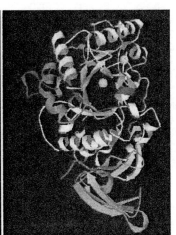

Figure 1. The CGTases share similarities with the alpha-amylases. The ribbon images of the proteins 1CDG and 6TAA were captured during an interactive 3D VRML session at the Protein Data Bank web site: http://www.rcsb.org/pdb/.

Molecular data for microbial proteins are readily accessible to anyone with a computer, but how do we use these resources? Visualization and bioinformatics tools prove useful as we take a closer look at microbial enzymes.

Part 1: Viewing Structural Models and Sequence Data

Although the two proteins above are clearly different molecules, we can see some similarity in the structure of the cyclomaltodextrin glucanotransferase (CGTase) and the alpha-amylase in Figure 1. Each of these enzymes contains a cylindrical cavity supported by alpha-helices (ribbon spirals) connected to beta-sheets (ribbon arrows) lining the interior of the cavity. Functionally similar, both proteins are active in the breakdown of starch molecules although the CGTases produce cyclomaltodextrins instead of the maltodextrins produced by alpha-amylases.

To examine the structure of the proteins above more closely, access the Protein Data Bank site at http://www.rcsb.org/pdb/.

Enter 1CDG into the box under Search the Archive and hit Find a structure. The Query Result Browser will appear. Click on 1CDG from the list of proteins to see the Summary Information page for 1CDG. (See Figure 2.)

Choose View Structure on top of the left column.

The View Structure page provides a list of options. Explore 1CDG and then 6TAA using at least the two options below:

1. To interactively explore the ribbon diagram above, click on VRML (default options) on top of the list. A 3D structural model will appear. To change the model's size, click on the Walk and move the cursor up or down. To rotate the model freely, click on Study.

Figure 2. Screen shots from the Protein Data Bank featuring the archive search and interactive 3D display options.

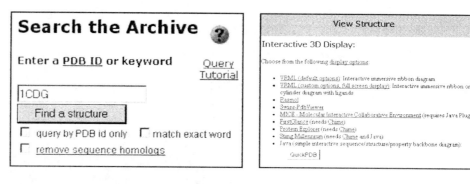

2. To explore the relationship between a wireframe structure of 1CDG and its protein sequence data, click on the QuickPDB at the bottom. You can rotate the model to view interesting structural components. Note the small Cursor window on the left. Individual amino acids and their position show up as you move over the model. Click on an amino acid in the long Sequence window across the top to highlight it in the model below.

Many proteins have been sequenced, but most lack the x-ray crystallography data necessary to generate structural models. Researchers interested in these proteins use the sequence data itself. Just as it takes time to learn to see information in a structural model, viewing sequence data meaningfully requires new analytical approaches and strategies.

Consider the sequence data for these two proteins in Tables 1 and 2. Asterisks mark the amino acid residues corresponding to an active site found in 1CDG at positions 225 - 233 and in 6TAA at positions 202 – 210. These conserved sequences confer specific properties to an active site and were retained during the evolution of these two proteins.

- Compare the conserved regions marked by asterisks in these two sequences. Which amino acids are the same?

Use the conserved sequence information in Table 3 to enter your own asterisks under the corresponding region of the alpha-amylase sequence in Table 2. (Note: The numbers beginning each row in Table 2. indicate the sequence position for the first amino acid residue in the row.)

Table 3: Well-conserved regions of alpha-amylase from *Aspergillus oryzae* with locations of specific amino acids and known structure-function relationships. (adapted from MacGregor 1996)

Starting and ending position of residues	Conserved sequence (Highly conserved in **bold**)	Some known structure - function relationships	
116 - 122	**V**DV**VANH**	Subsite 1	H = His 122
202 - 210	G**LRID**TV**KH**	Catalytic site Subsite 1 Subsite 2	D = Asp 206 H = His 210 K = Lys 209
230 - 233	**EVLD**	Catalytic site	E = Glu 230
293 - 298	V**ENH**DN	Catalytic site Subsite 1	D = Asp 297 H = His 296

1	APDTSVSNKQ	NFSTDVIYQI	FTDRFSDGNP	ANNPTGAAFD	GTCTNLRLYC
51	GGDWQGIINK	INDGYLTGMG	VTAIWISQPV	ENIYSIINYS	GVNNTAYHGY
101	WARDFKKTNP	AYGTIADFQN	LIAAAHAKNI	KVIIDFAPNH	TSPASSDQPS
151	FAENGRLYDN	GTLLGGYTND	TQNLFHHNGG	TDFSTTENGI	YKNLYDLADL
201	NHNNSTVDVY	LKDAIKMWLD	LGIDGIRMDA	VKHMPFGWQK	SFMAAVNNYK
			* * * * * *	* * *	
251	PVFTFGEWFL	GVNEVSPENH	KFANESGMSL	LDFRFAQKVR	QVFRDNTDNM
301	YGLKAMLEGS	AADYAQVDDQ	VTFIDNHDME	RFHASNANRR	KLEQALAFTL
351	TSRGVPAIYY	GTEQYMSGGT	DPDNRARIPS	FSTSTTAYQV	IQKLAPLRKC
401	NPAIAYGSTQ	ERWINNDVLI	YERKFGSNVA	VVAVNRNLNA	PASISGLVTS
451	LPQGSYNDVL	GGLLNGNTLS	VGSGGAASNF	TLAAGGTAVW	QYTAATATPT
501	IGHVGPMMAK	PGVTITIDGR	GFGSSKGTVY	FGTTAVSGAD	ITSWEDTQIK
551	VKIPAVAGGN	YNIKVANAAG	TASNVYDNFE	VLSGDQVSVR	FVVNNATTAL
601	GQNVYLTGSV	SELGNWDPAK	AIGPMYNQVV	YQYPNWYYDV	SVPAGKTIEF
651	KFLKKQGSTV	TWEGGSNHTF	TAPSSGTATI	NVNWQP	

Table 1. Sequence Information for 1CDG, a CGTase produced by a bacterium. From the Protein Data Bank: www.rcsb.org/pdb/.

1	ATPADWRSQS	IYFLLTDRFA	RTDGSTTATC	NTADQKYCGG	TWQGIIDKLD
51	YIQGMGFTAI	WITPVTAQLP	QTTAYGDAYH	GYWQQDIYSL	NENYGTADDL
101	KALSSALHER	GMYLMVDVVA	NHMGYDGAGS	SVDYSVFKPF	SSQDYFHPFC
151	FIQNYEDQTQ	VEDCWLGDNT	VSLPDLDTTK	DVVKNEWYDW	VGSLVSNYSI
201	DGLRIDTVKH	VQKDFWPGYN	KAAGVYCIGE	VLDGDPAYTC	PYQNVMDGVL
	* * * * * * * * *				
251	NYPIYYPLLN	AFKSTSGSMD	DLYNMINTVK	SDCPDSTLLG	TFVENHDNPR
301	FASYTNDIAL	AKNVAAFIIL	NDGIPIIYAG	QEQHYAGGND	PANREATWLS
351	GYPTDSELYK	LIASANAIRN	YAISKDTGFV	TYKNWPIYKD	DTTIAMRKGT
401	DGSQIVTILS	NKGASGDSYT	LSLSGAGYTA	GQQLTEVIGC	TTVTVGSDGN
451	VPVPMAGGLP	RVLYPTEKLA	GSKICSSS		

Table 2. Sequence Information for 6TAA, an alpha-amylase (also known as TAKA amylase) found in *Aspergillus oryzae*. From the Protein Data Bank: www.rcsb.org/pdb/.

Now that the conserved sequence VDVVANH is marked by asterisks in Table 2, try to locate the corresponding sequence for CGTase in Table 1. Enter asterisks appropriately in Table 1. (Note: This is not easy to do. Imagine trying to manually compare hundreds of proteins for multiple sites! Part 2. introduces bioinformatics tools that will align and identify conserved sequences in two or more proteins.)

- List the starting and ending positions and the amino acid sequence you've found in 1CDG that correspond to this active site in 6TAA.

- Which amino acids are conserved?

Variations in the amino acids can change the functionality of the active site. For example, CGTase sequences have a phenylalanine residue located near the active site that corresponds to the glutamic acid residue (E230) in the alpha-amylase sequence EVLD shown in Table 3. The phenylalanine is required for cyclization of the cleaved maltodextrin fragments. (MacGregor, 1993)

Part 2: Using Sequence Data with Online Bioinformatics Tools

The sequence information for one CGTase can be used to probe for similar proteins in an online molecular database. (See "Orientation to the *Biology Workbench*" on the *Microbes Count!* web site and the "Searching for Amylase" activity in Chapter 2 for an explanation of how to get started.) In *Biology Workbench*, use Ndjinn to search for the 1CDG protein in the PDBSEQRES database.

- View the record. What organism produces this protein?

Import the sequence, and use BLASTP to find similar proteins in the SWISS-PROT database.

- Choose three of the CGTases that are displayed, but make sure each is produced by a different organism. List the three CGTases and the organism that produces them.

Use CLUSTALW to align the sequences for these three proteins with the 1CDG protein. Consider the output of aligned sequences. Are you surprised by the number of amino acid residues that are conserved?

- CLUSTALW also produces an unrooted tree from the sequence data. Usually the tree is near the bottom of the output file. Use the tree to decide which of your CGTases are the most similar to the 1CDG protein. Defend your choice.

Not all CGTases are used in industrial starch processing. A CGTase used to make alpha-cyclodextrins commercially is produced by *Paenibacillus macerans*, formerly known as *Bacillus macerans* (Guzman-Maldonado and Paredes-Lopez, 1995). The SWISS-PROT label for this CGTase is CDG1_PAEMA.

- Devise a strategy using sequence information to determine which of the CGTases you are working with (including 1CDG) would be most likely to function similarly in the breakdown of starch. Explain your results.

Part 3. An Extended Research Problem: Making the Case for a New Microbial Enzyme

- The protein AMYR_BACS8 is produced by a microbe that can digest raw starch. Is this advantageous for the *Bacillus*? Why?

- Would this be of value to humans? Explain.

- You have been asked to present a report at the planning meeting for a starch processing plant on possible applications of this microbial enzyme. Your role is to provide information on the protein itself. Identify three questions you anticipate will be asked during the meeting.

- Prepare a short presentation that both the research scientists and the marketing specialists will understand. In addition to looking for information from publications on the protein, present the available molecular data drawing from your experience with the visualization and bioinformatics techniques in this activity.

Web Resource Used in this Activity

Biology Workbench (http://workbench.sdsc.edu)
Originally developed by the Computational Biology Group at the National Center for Supercomputing Applications at the University of Illinois at Urbana-Champaign. Ongoing development of version 3.2 is occuring at the San Diego Supercomputer Center, at the University of California, San Diego. The development was and is directed by Professor Shankar Subramaniam.

Protein Data Bank
http://www.rcsb.org/pdb/

VRML Plug-in: Cosmo Player
You will need the Cosmo Player plug-in to use VRML. See download information on the PDB site or http://www.karmanaut.com/cosmo/player/

Additional Resources

Available on the *Microbes Count!* web site at http://bioquest.org/microbescount

Text

A copy of this activity, formatted for printing

"Orientation to the *Biology Workbench*"

Related *Microbes Count!* Activities

Chapter 2: Searching for Amylase

Chapter 4: Molecular Forensics

Chapter 4: Exploring HIV Evolution: An Opportunity for Research

Chapter 6: Proteins: Historians of Life on Earth

Chapter 6: Tree of Life: Introduction to Microbial Phylogeny

Chapter 6: Tracking the West Nile Virus

Chapter 6: One Cell, Three Genomes

Unseen Life on Earth Telecourse

Coordinates with Video VII: Microbial Diversity

Relevant Textbook Keywords

Active site, Amino Acid, Bioinformatics, Conserved sequence, Molecular model, Visualization, X-ray crystallography,

Related Web Sites (accessed on 4/23/03)

Examining protein structures
http://www.ornl.gov/TechResources/Human_Genome/posters/chromosome/pdb.html

Microbes Count! Website
http://bioquest.org/microbescount

Molecular Graphics Manifesto
http://www.usm.maine.edu/~rhodes/Manifesto/text/01Intro.html

Tutorials on How to Use RasMol and Chime
http://www.umass.edu/microbio/rasmol/rastut.htm

Unseen Life on Earth: A Telecourse
http://www.microbeworld.org/htm/mam/is_telecourse.htm

Bibliography

Guzman-Maldonado, H. and O. Paredes-Lopez (1995). Amylolytic enzymes and products derived from starch: A review. *Critical Reviews in Food Science and Nutrition.* 35 (5):373-403.

Janacek, S., E. Leveque, A. Belarbi, and B. Haye (1999). Close evolutionary relatedness of alpha-amylases from archaea and plants. *J. Molecular Evolution* 48:421-426.

MacGregor, E. A. (1988). a-Amylase structure and activity. *J. Protein. Chem.* 7:399-415.

MacGregor, E. A. (1993). Relationships between structure and activity in the alpha-amylase family of starch-metabolising enzymes. *Starke* 45:232-237.

MacGregor, E. A. (1996). Structure and activity of some starch metabolizing enzymes in *Enzymes for Carbohydrate Engineering*. Park, K-H., J. F. Robyt, and Y. D. Choi, Editors. Elsevier: Amsterdam.

Figure and Table References

Figure 1.   Courtesy Sam Donovan (Beloit College)

Figure 2.   Modified from *Biology WorkBench* (http://workbench.sdsc.edu)

Table 1.   From the Protein Data Bank: www.rcsb.org/pdb/

Table 2.   From the Protein Data Bank: www.rcsb.org/pdb/

Table 3.   Adapted from MacGregor 1996

Chapter 8

Activities for Video VIII: Microbial Ecology

Current attempts to understand microbial ecology have dramatically changed the way we do research. No longer are we content to focus on the microbes we can grow in pure culture on carefully determined media, but now we seek naturally occurring microbes within the complex situations they successfully negotiate in the environment.

In this unit we can:

- study the effects that the presence of a refuge from predators has on a model microbial population, and
- measure key metabolic indicators and observe succession in a fermentation culture for making Kimchee.

Modeling Microbial Predator-Prey Relationships

In a mixed culture, how does the population of Didinium affect the population of Paramecium – and vice versa? Predator-prey cycles can be modeled using the Biota simulation. How do variables such as the presence of a refuge or the availability of food perturb the population cycles?

Exploring Microbial Fermentation with Korean Kimchee

To get a closer look at metabolism and succession, consider the fermentation process for making traditional Kimchee. This non-pathogenic system provides opportunities for real time data acquisition and analysis. Experimental data is provided for variables such as pH and O2 levels.

Investigating Predator-Prey Interactions

Ethel Stanley

Video VIII: Microbial Interactions

It's a jungle out there… even for a microbe in a drop of water! Using a modeling program called *Biota*, you can look more closely at the predator-prey relationship between *Didinium* and *Paramecium*. These two microorganisms are commonly found in pond water and have been studied since the early 20th century.

Getting Started

The Biota application is located in the Predator-Prey folder on the Microbes Count! web site. For additional information on using Biota, please consult the User's Manual and Biota tutorial, also located on the site.

Double-click the Biota icon to see a startup box. Choose Open Document and then choose the simulation file named **Didinium-Paramecium**. Two windows will open, Field Notes and Didinium-Paramecium map. You may wish to read about the model in the Field Notes, but move this window to see the model. The Map shows a single region, A1, representing a tank that provides an aquatic habitat for both microorganisms.

To see the interactions between the model components, choose Species on the menubar and highlight Interactions from the pop-up menu below it. The Species Interaction Window (see Figure 2) will open. Click on any cell in the table to see the effect the model components have on each other. Do not make changes to the values at this time, but you may choose to do so later when you are constructing your own models. Click OK to close the window.

The Interaction window provides a description of all the species interactions. Click once on a cell in the table at the bottom to display the information. The three boxes highlighted here include *Paramecium* (prey), *Didinium* (predator), and the explanation of the positive value *Paramecium* contributes to *Didinium*. (Note: Ignore the column for Refuge. Its effects will be investigated later.)

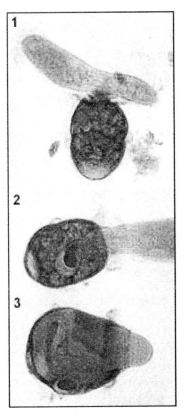

Figure 1. *Didinium* is a voracious predator. In this series, a *Paramecium* is attacked and ingested in a few minutes.

Figure 2. This window provides a description of all the species interactions.

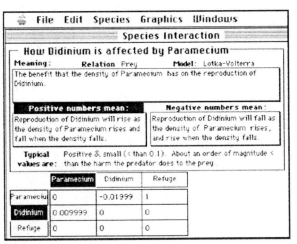

Choose Start Simulation… under Species to observe the predator-prey relationship when there is no escape for the prey. In the simulation summary window that appears, use the Step button to move through 25 cycles.

- How does the Didinium population change over time?

Revise the *Didinium-Paramecium* Model

Double the initial population of the prey and run the simulation again. Double click Paramecium in the A1 region. Enter 100 to replace the 50 shown as the starting density in the Deme Info window (see Figure 3). Run the simulation again for 25 cycles.

Figure 3. You can change the starting density of the *Paramecium* population by entering new values in the Deme Info window.

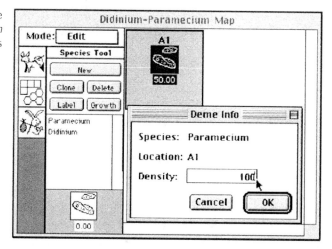

- How does the change in the *Paramecium* population impact the predator? If the *Paramecium* population were doubled again, what would you expect to see? Revise the model and observe what happens.

- What would you expect to happen if you doubled the initial predator population?

- Revise the model to double the predator population. Do the results support your prediction? Explain. (Hint: Reset the model to 50 for the *Paramecium*. Then click once on *Didinium* under Species Tool. When the icon for *Didinium* appears in the A1 region, double click it to get the Deme Info window. Make your changes and Run the simulation.

Like you, an investigator named Gause found that there are cycles in which the *Didinium* population grows as they eat the *Paramecium*. In the lab, *Didinium* quickly drive the *Paramecium* to extinction and then the *Didinium* themselves die from starvation. This led Gause to consider how both *Didinium* and *Paramecium* remain viable in nature. In 1934, Gause hypothesized that if the Paramecium had a refuge from predation, then they could avoid extinction. In his lab, he added a bit of glass wool to provide cover for the *Paramecium*, preventing successful hunting by the *Didinium*.

Explore the Refuge Model

Open the file named Refuge. In this simulation, each refuge in the "tank" provides protection for one *Paramecium*. Run this simulation.

- Why do ecologists refer to the predator-prey relationship as a cycle?

- Does the refuge make a difference? Explain.

- Consider the likelihood that refuges enable fast growing microbes to survive not only predation, but also sanitation efforts. List at least five refuges bacteria have in your kitchen.

Explore the Food Model

Choose the file Food. In this simulation, food for the *Paramecium* is constantly provided. Run this simulation for at least 200 cycles.

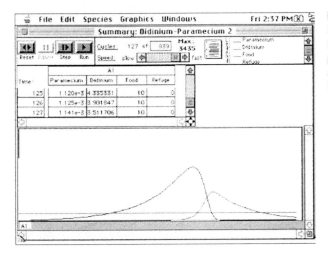

Figure 4. Note the initial increase in the *Paramecium* population followed by a rise in the *Didinium* population. The constant food availability is shown by the straight line.

- What effects does food for the *Paramecium* have on the *Paramecium* population?

- What effects does food for the *Paramecium* have on the *Didinium* population?

Run the Bacteria Model

The Food Model is somewhat unrealistic, since the sources of food do not usually remain constant. Close this model and open the Bacteria Model. Here the model introduces bacteria that are eaten by the *Paramecia*. The new species, Bacteria, has both growth characteristics and interaction with other species.

- What kind of population results do you get with the Bacteria model?

You can use Biota to build models to further explore interactions between species and their effect on predator-prey relationships. Are predator and prey always coupled by cyclic population growth? Can predator-prey populations recover in disturbed situations? If you run a simulation longer, do you see increased stability? Investigate a problem of interest to you or your group by constructing your own model.

- How will you change the model to explore this question?

- What do you think will happen when you run the model? Write down your hypothesis.

- Run the revised model. Did your results support your hypothesis? Explain.

Optional Activity

In a study of microbial food webs in the coastal waters within Onagawa Bay on the northeastern Pacific coast of Japan, Tanaka (1998) reported " microbial eddies" between the heterotrophic nanoflagellates (HNF) and bacterioplankton. Increases in bacterial abundance were usually followed by increases of HNF abundance.. The level of HNF varied between 30 - 6,700 cells/ml during the year.

- What is the likely nature of the relationship between these two populations?

- Offer an explanation for this observed variation in population densities for both the bacterioplankton and the HNF. (Note: Consider factors you've modeled as well as environmental factors such as the geographic location of these populations.)

- Explore current literature on microbial food webs and provide at least two examples of microbial predator prey interactions.

Software Used in this Activity

Biota

Jim Danbury, Ben Jones, John Kruper, Eric Nelson, William Sterner, Jeff Schank, Jim Lichtenstein, Joyce Weil, and William Wimsatt (University of Chicago)

Platform Compatibility: Macintosh only

Additional Resources

Available on the *Microbes Count!* web site at http://bioquest.org/microbescount

Software

Biota

Text

A PDF copy of this activity, formatted for printing

Biota User's Manual and Biota Tutorial

Related *Microbes Count!* Activities

Chapter 3:   Modeling Microbial Growth

Chapter 9:   Mold Fights Back: A Challenge for Sanitation

Chapter 9:   A Plague on Both Houses: Modeling Viral Infections to Control
             a Pest Outbreak

Unseen Life on Earth Telecourse

Coordinates with Video VIII: Microbial Ecology

Relevant Textbook Keywords

Didinium, Food web, *Paramecium*, Refuge

Related Web Sites (accessed on 4/14/03)

Data set: Prey species: Paramecium aurelia / Predator species: Didinium nasutum
    http://www.inapg.inra.fr/ens_rech/bio/Ecologie/fichiers/PRSLBDataSets.txt

Images/Movie: *Didinium* feeding on *Paramecium*
    http://www.microscopy-uk.org.uk/mag/art97/dingley3.html
    http://micro.magnet.fsu.edu/moviegallery/pondscum/protozoa/didinium/index.html

Microbes Count! Website
    http://bioquest.org/microbescount

Predator-prey interactions in nature and in the laboratory
    http://viceroy.eeb.uconn.edu/eeb244site/Predation/02-15_PredationDynamics.html

Predator Prey Dynamics
    http://www.biology.ualberta.ca/courses.hp/bio331.hp/lectures/lect22/PredatorPreyDynamics.htm
    http://cas.bellarmine.edu/tietjen/Pop_Models/OCX/Prey/PredatorTxt.htm

Summary of Gause's 1934 studies of *Paramecium-Didinium* predation and refuge study
    http://www.idlex.freeserve.co.uk/idle/evolution/ref/pianka1.html

Unseen Life on Earth: A Telecourse
    http://www.microbeworld.org/htm/mam/is_telecourse.htm

References

Gause, G.F. (1934). *The Struggle for Existence*. Hafner, New York

Danbury, J., B. Jones, J. Kruper, E. Nelson, W. Sterner, J. Schank, J. Lichtenstein, J. Weil, W. Wimsatt (2001). Biota. In *The BioQUEST Library Volume VI*, Jungck, J. R. and V. G. Vaughan, Editors. Academic Press: San Diego, California.

Tanaka, T. (1998) Dynamics of the heterotrophic nanoflagellate-bacteria associations in a coastal marine environment
http://www.aslo.org/phd/dialog/1998January-35.html

Figure and Table References

Figure 1. Courtesy Michael Dingley, Australia
http://www.microscopy-uk.org.uk/mag/art97/dingley3.html

Figure 2. Screenshot from *Biota*

Figure 3. Screenshot from *Biota*

Figure 4. Screenshot from *Biota*

Exploring Microbial Fermentation with Korean Kimchee

John M. Greenler and Robin McC. Greenler

Video VIII: Microbial Ecology

Introduction

The goal of this activity is to investigate microbial fermentation and ecology. Students will monitor and quantify the population growth and metabolism in a self-designed experiment centered on the production of kimchee. Kimchee is a traditional Korean food that is the product of fermentation of napa cabbage.

Figure 1. There are hundreds of different types of kimchee.

Fermentation

Fermentation is the process through which microbes metabolize organic compounds under anaerobic conditions to yield energy. It can result in the production of a number of different chemicals including ethanol (see "Modeling Wine Fermentation" in Chapter 3) or, as in the case of kimchee, lactic acid. Both lactic acid and alcoholic fermentation have formed the basis for a significant part of the human diet for thousands of years.

By controlling the fermentation process through starting substrate, microbial innoculum, and environmental conditions, the course of fermentation can result in a wide variety of final products. Fermented milk products include yogurt, cheese and buttermilk, and fermented fruits and grains can result in alcoholic beverages, sourdough bread and fermented soy products.

Microbially fermented vegetables are common to many cultures including German sauerkraut, Japanese tsukemono, Nepalese gundruk as well as Korean kimchee. While kimchee has unique characteristics and is a keystone of Korean culture, it has features common to all microbially fermented vegetables.

Background

Records of Kimchee production and consumption date back to the 3rd or 4th century (Man-Jo et al. 1999). There are hundreds of kimchee recipes that are tailored for every meal, various times of the year, and different social occasions (Lee et al. 1998). In 1994, it was estimated that kimchee made up 12% of the average diet in Korea. (Cheigh and Park, 1994).

In its most basic form, kimchee is made by salting napa cabbage (*Brassica napa*, also called Chinese or Korean cabbage), and letting it ferment. However, Kimchee generally includes a variety of other vegetables, seafood and seasonings. The components can directly affect the microbial populations by introducing new microbial populations or ingredients with bactericidal properties such as garlic and pepper. The additional ingredients can also provide limiting nutrients that enhance bacterial growth such as the nitrogen contained in seafood. The starting material and microbes involved then interact with the environmental conditions to promote a specific ecological succession of microbes throughout the fermentation process that ultimately leads to a final product with a distinctive flavor and texture.

Microbial Succession

In the process of fermentation, different microbes are active at different pH levels. Higher concentrations of salt and lower temperatures may slow fermentation. At the earliest stages of fermentation, aerobic bacteria and anaerobic bacteria co-exist. Lactic acid, acetic acid and carbon dioxide are produced by bacterial metabolism of sugars. As salt concentrations initiate osmosis, the released liquid results in a more anaerobic environment, although aerobic bacteria, molds and yeasts, continue to grow at surface layers.

Around 200 species of microbes can be found in kimchee at different stages of the fermentation process. *Leuconostoc mesenteroides* is frequently a dominant bacterium during the initial, predominately aerobic, stages. This species then decreases as the pH drops around day four to six and *Lactobacilli* species increase, as shown in Figure 2 (Cheigh and Park, 1994). Figure 3 illustrates a rise in total acids and a drop in reducing sugars concomitant with the drop in pH (Cheigh and Park, 1994).

Figure 2 (left): Bacterial succession in kimchee during fermentation at 20 °C.

Figure 3 (right): Changes in total acids, reducing sugars, and pH in kimchee during fermentation at 20°C (salt 3%).

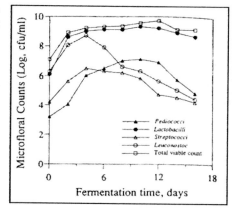
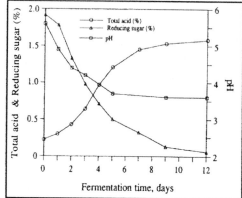

Investigating Microbial Fermentation with Kimchee

Working in teams, students design and carry out a research experiment based on manipulation of one or more variables that might affect the fermentation of kimchee. Students monitor the changes that occur during fermentation using methods ranging from pH paper to real time data acquisition sensors. Because fermentation studies can use materials available in the grocery store, these experiments can be done safely by distance-learning students at home. (Christina Strickland, Clackamas Community College, personal communication).

With a kimchee fermentation chamber, students can observe and quantify the course and rate of fermentation, the successional sequence of the microbes, the microbial populations and the final product. They can consider the role of ambient conditions, starting ingredients, and preparation of those ingredients. They can compare traditional practices and recipes with respect to the microbial ecology and course of fermentation. Over the period of about a week, the changes that occur can be monitored through visual and olfactory observations as well as measurement of factors such as pH, turbidity, rate of CO_2 production and O_2 consumption.

Student teams must first form a research question based on consideration of one or more variables that might affect the fermentation of kimchee. Often this process is more successful if the students are allowed to observe an initial fermentation run as a way to familiarize themselves with the lab techniques and to generate questions. Depending on the research question they are asking, students must decide how and what they want to monitor the during the fermentation process.

Building a Kimchee Fermentation Chamber

A satisfactory kimchee fermentation chamber can be constructed using 2-liter plastic soda bottles (Williams et al. 1993). In addition to being inexpensive and accessible, these bottles allow individual students to design experiments with multiple replicates, several runs, or modified chamber design. The instructions below represent a starting point for construction of a standard kimchee fermentation chamber and initiation of a kimchee fermentation time course. Construction and filling of the chamber may modified to meet the students' question and experimental design.

Materials

- One 2-liter soda bottle
- One plastic lid about 9 cm diameter (a petri dish lid can be used)
- 1 head of Chinese cabbage, roughly cut (approximately 1, 000 grams)
- Non-iodized, pickling or kosher salt equal to 2% of cabbage weight
- Scissors
- Safety razor or utility knife
- Wax pencil or marker
- Two cloves of garlic (optional)
- Chopped hot chili peppers or chili powder (optional)

Construction of Chamber

1. Remove the label from a 2-liter soda bottle. As the glue that adheres the label is heat sensitive, running hot tap water inside or outside the bottle on the glue line will allow easy removal.

2. Cut the top off the bottle just 1 cm below the shoulder of the bottle. (The shoulder is the place where the tapered top of the bottle meets the straight sides.)

3. In order to cut the bottle, first draw a cut line around it by placing the bottle on its side with the base against a wall or vertical surface and, while holding the pen tip to the bottle, slowly turn the bottle thus drawing a line around it. Once the line is drawn, begin the cut with a safety razor or utility knife and

continue it with a scissors. It is easier to cut the bottle with the top arm of the scissors inside the bottle.

Preparation of Fermentation Experiment

1. Weigh cabbage to be used and determine the amount of salt needed. The standard salt concentration is 2% of cabbage dry weight, however, students may choose to use different concentrations to monitor effects.

2. Occasionally, the initial microbial populations are modified by washing the cabbage briefly in a 10% bleach solution (and then rinsing well) or by "brining" (soaking) the cabbage in a concentrated salt solution (3-5%) for 1-15 hours.

3. Layer cabbage, salt and spices (optional) into the bottle until completely full, packing the mixture very firmly to break up the cabbage cells.

4. Insert the plastic lid onto the cabbage and place a weight on the lid to keep the mixture submerged in the liquid that will emerge in the first few hours.

5. After a few hours, the cabbage will fill 1/2 to 2/3 of the bottle. Slide the bottle top inside the chamber, push down so no air remains under the plastic lid and straighten the bottle top to make an airtight seal. The cabbage should be submerged at all times. (Note: Leave the screw cap on the bottle lid, but do not tighten it all of the way down. By leaving the cap loose, gasses produced during fermentation can vent out of the chamber, but still prevent significant amounts of oxygen from entering.)

Data Acquisition during Kimchee Fermentation

The process of microbial fermentation in kimchee can be monitored in numerous ways using a wide range of technologies. By using sensors interfaced to a computer it is possible to collect real-time data and easily perform subsequent computer-based analysis (Figure 4). The Vernier Software and Technology Company (www.vernier.com) provides a wide range of sensors along with interfaces and software that are economical and appropriate for undergraduate use.

Below is a list of some of the variables that can be monitored during the fermentation of kimchee. Depending on the question posed and the experimental design, any combination of these or other data acquisition strategies will be used.

- pH: Acidity can be monitored in numerous ways. A pH paper strip will give quick and easy results. Using a pH electrode will give more accurate results and in some cases can be interfaced directly with a computer.

- Microbial population size: Use a serial dilution technique to estimate bacterial counts.

- Turbidity: As fermentation occurs, the turbidity of the solution increases. This is due at least to the increase in microbial populations as well as breakdown of the starting ingredients. Turbidity can be measured with a turbidity sensor. (Note: After reaching a turbidity reading of 100 NTUs, it is necessary to dilute the sample 1:10 with a salt solution of the same concentration to get an accurate reading. This dilution must then be compensated for in subsequent calculations). It should be noted that when measuring turbidity, both live and dead microbes contribute to the overall turbidity.

- CO_2 production: CO_2 is a by-product of kimchee fermentation. It can be measured with a CO_2 gas sensor. This can be done by removing the screw-cap from the lid and replacing it with a cap that has had a hole drilled in it just big enough for a CO_2 sensor to slide through. (Note: It is easiest to measure the rate of CO_2 production over a relatively short period of time, such as five minutes. To do this flush out the existing air in the chamber before beginning measurements. This flushing can be achieved with an air line or a simple pump such as those used for sports balls.)

- O_2 consumption: Because there will always be some aerobic microbes present during the production of kimchee, there will be oxygen consumption. This can be measured via an oxygen sensor in a manner similar to that used for the measurement of CO_2.

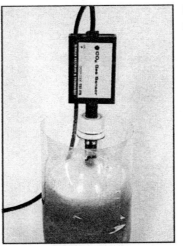

Figure 4: A Vernier CO2 gas sensor measuring Kimchee fermentation

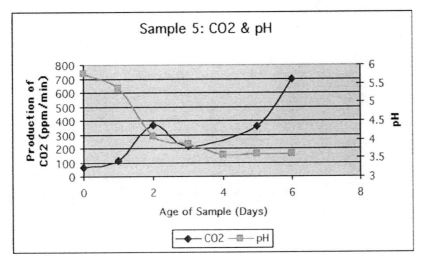

Figure 5: Sample data from kimchee fermentation at 20 °C, 2% salt

Questions

1. What limits bacterial growth? Are there factors that stop the growth? How will you evaluate bacterial growth? Do your answers vary for different species of microbes?

2. How do turbidity and the rate of CO_2 production differ as measures of bacterial growth?

3. How does washing the cabbage in a 10% bleach solution (and then rinsing well) prior to fermentation affect the ensuing process or the final product?

4. Brining kimchee may remove some microorganisms initially as well as beginning the osmosis of water from cabbage. How, if at all, does this affect the course of fermentation or the final product?

5. Plate bacteria on lactobacilli agar or nutrient agar plates to examine colony morphology and identify microbes.

In-Depth Challenge

Characterize the microbial succession. Examine variables that alter the successional sequence or rate.

Further Thinking Activities

Consider the phase diagrams from several fermentation runs of kimchee (shown on the next page). The kimchee was fermented at 20 degrees C, with a 2% salt solution. Raw data for these and other runs are included on the *Microbes Count!* web site.

- Biologically, what is the nature of the relationships?

- What phases of the growth cycle are represented in various points of each graph?

- During the course of fermentation, what major acid metabolites are accumulating?

- How might the metabolites affect other parameters?

- Based on the pH v O_2 and O_2 v CO_2 graphs, what can we tell about which populations of microbes (aerobic, anaerobic or facultative) might be dominant in which phase?

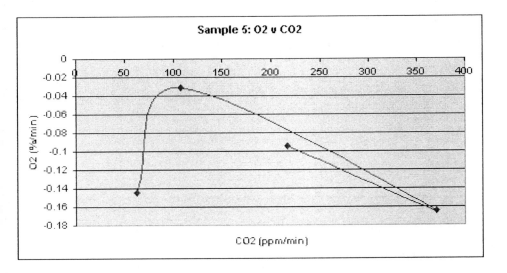

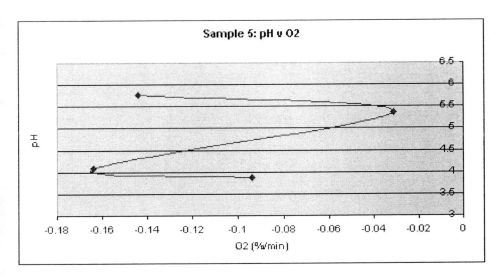

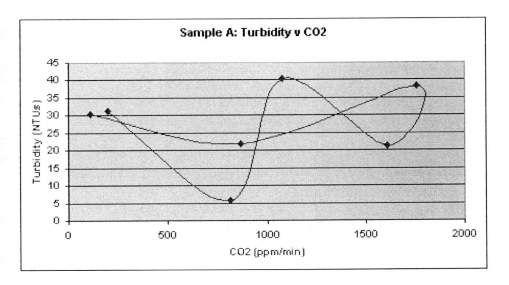

Additional Resources

Available on the *Microbes Count!* web site at http://bioquest.org/microbescount

Text

A PDF copy of this activity, formatted for printing

Data

Data from kimchee fermentation experiments

Related *Microbes Count!* Activities

Chapter 3: Modeling Wine Fermentation

Chapter 3: The Living World of Yogurt

Unseen Life on Earth Telecourse

Coordinates with Video VIII: Microbial Ecology

Related Web Sites

Korean Food Research Institute
http://kimchi.kfri.re.kr/

Microbes Count! Website
http://bioquest.org/microbescount

Unseen Life on Earth: A Telecourse
http://www.microbeworld.org/htm/mam/is_telecourse.htm

Vernier Software and Technology Company
http://www.vernier.com

References

Cheigh, H. and K. Park (1994). Biochemical, microbiological, and nutritional aspects of Kimchi (Korean fermented vegetable products). *Critical Reviews in Food Science and Nutrition* 34(2):175-203.

Korea Food Research Institute website, http://kimchi.kfri.re.kr/

Lee, C. J., H. W. Park and K. Y. Kim (1998). *The Book of Kimchi*. Korea Information Service: Seoul, The Republic of South Korea.

Man-Jo, K., Kyou-Tar, L., O-Young, Lee (1999). *The Kimchee Cookbook, Fiery Flavors and Cultural History of Korea's National Dish*. Periplus: Boston.

Williams, P., J. Greenler, R. Greenler, L .Graham, M. Ingram, L. Kehle and D. Eagan (1993). *Bottle Biology, an Idea Book for Exploring the World through Soda Bottles and Other Recyclable Materials*. Kendall/Hunt Publishing Co.: Dubuque, Iowa.

Yip, D. Y. (2000). Promoting a better understanding of lactic acid fermentation. *Journal of Biological Education* 35(1):37-40.

Figure and Table References

Figure 1:   John Greenler

Figure 2:   Modified from Cheigh and Park, 1994

Figure 3:   Modified from Cheigh and Park, 1994

Figure 4:   John Greenler

All other data graphs courtesy of Tia Johnson.

Chapter 9

Activities for Video IX: Microbial Control

Issues of microbial control extend beyond the traditional focus on the patient and the hospital to the complexities of microbial control in the environment. Simulations of real world problems engage us in thinking through these complexities including bioterrorism, an all too familiar reality of our modern global society.

In this unit we can:

• develop a plan to decontaminate areas in the Senate Office Building where anthrax spores have been found,
• investigate the efficacy of different strategies to control mold growth on a virtual hospital floor,
• examine international alternatives for the control of citrus canker, and
• explore the dynamics of the evolving Myxoma virus on virtual rabbit populations in an Excel workbook.

Bioterror and the Microbiologist's Response

The discovery of anthrax spores in the Senate Office Building in Washington in October 2001 as well as in several post offices caught microbiologists unprepared for their central role in the War on Terrorism. Infection control specialists were challenged to think of new strategies to confront the anthrax problem. How would you respond to this crisis?

Mold Fights Back: A Challenge for Sanitation

What do we mean by control? Using an actual incident involving a broken water main and rapid mold growth in a new hospital as the basis for the model, this EcoBeaker simulation offers an opportunity to set up and then test control methods. In a second model, the emergence of resistant mold raises new challenges.

Citrus Canker: Alternatives for Control

What do you do when county officials show up to cut down the orange trees in your backyard? What causes citrus canker and how is it spread? This plant pathogen was the first microbe to have its genome sequenced outside of the US. There is much to investigate before deciding on the best alternative for control.

A Plague on Both Houses: Modeling Viral Infection to Control a Pest Invasion

Introduced as a control measure for reducing the rabbit population in Australia, the Myxoma virus produced some unintended effects. A computationally rich Excel workbook based on a SIR model for disease transmission provides an opportunity to explore the use of an infectious disease as a control agent. Modeling the emergence of multiple Myxoma strains offers intriguing insights into the complexities of control.

Bioterror and the Microbiologist's Response

Marion Field Fass

Video IX: Microbial Control

The goal of this activity is to apply information on methods of sterilization to the problem presented by the discovery of anthrax *(Bacillus anthracis)* spores in a large, well-utilized building. This activity provides the opportunity to review standard and non-standard strategies for disinfection and sterilization and to determine the optimal methods for cleaning up a substance whose characteristics are not fully understood.

Starting in October 2001, microbiologists were asked to step into roles that they had only imagined. When anthrax began to arrive in letters sent to media and to the Hart Senate Office Building, microbiologists were among the first called in to determine how to clean up after a terrorist attack and how to judge when a building was again safe.

Figure 1. FBI and Environmental Protection Agency personnel working to identify anthrax-contaminated mail at the containment facility.

Public health officials and microbiologists had planned for attacks of biological weapons before the anthrax attacks of 2001, but their scenarios were inadequate. Scientists had developed strategies for responding to large-scale attacks, but had not prepared for the psychological impact of limited but threatening attacks.

Anthrax-laced letters sent to Senator Tom Daschle at the Hart Senate Office Building in Washington, DC presented a real challenge for microbiologists and investigators. After the building was evacuated and all of the potentially exposed workers had received antibiotics to prevent the establishment of infection, no one was sure how to disinfect the building to make it safe for the return of workers.

The project

You have been hired by the United States government to develop a plan to decontaminate the Hart Senate Office Building and the post office in Trenton, New Jersey after the arrival of anthrax-laced letters. Armed with your Microbiology textbook, you consult the chapter on Sterilization to plan your approach.

Consider these needs:

- The Senate Office Building contains valuable documents that cannot be destroyed. The post office still has millions of undelivered letters.

- Workers need to be able to return to the buildings, although they do not need to return immediately.

- You need to guarantee the safety of workers who return.

Take into account the nature of the biological weapon:

- What are the characteristics of anthrax spores that make disinfection more difficult than if an individual with smallpox had walked through the building?

- What are the uncertainties with which you must cope?

Develop a table that identifies major infection control strategies, and cite the advantages and disadvantages for each of them in cleaning up after anthrax.

- What strategy have you chosen as best? Why?

- Can you devise an experimental setting to test the effectiveness of your strategy?

- Would the strategy for dealing with the post office be any different from your stategy for the Hart Senate Office Building?

- What advice would you give to workers concerned with contaminating their family members?

- How would you advise people who have started microwaving their mail in order to decontaminate it?

Additional Resources

Available on the *Microbes Count!* web site at http://bioquest.org/microbescount

Text

A PDF copy of this activity, formatted for printing

Related *Microbes Count!* Activities

Chapter 1:   Scale of the Microbial World

Chapter 9:   Mold Fights Back: A Challenge for Sanitation

Chapter 9:   Citrus Canker: Alternatives to Control

Chapter 9:   A Plague on Both Houses: Modeling Viral Infection to Control a Pest Outbreak

Unseen Life on Earth Telecourse

Coordinates with Video IX: Microbial Control

Relevant Textbook Keywords

Anthrax, *Bacillus anthracis*, Biological weapons (warfare), Bioterrorism, Disinfection, Endospore, Spore, Sterilization methods

Related Web Sites (accessed on 2/20/03)

Centers for Disease Control and Prevention, Public Health Emergency
Preparedness and Response
http://www.bt.cdc.gov/agent/anthrax/index.asp

Microbes Count! Website
http://bioquest.org/microbescount

National Library of Medicine's Information on Biological Warfare
http://www.sis.nlm.nih.gov/Tox/biologicalwarfare.htm

Unseen Life on Earth: A Telecourse
http://www.microbeworld.org/htm/mam/is_telecourse.htm

U.S. Department of Labor, Occupational Safety and Health Administration,
ANTHRAX eTool
http://www.osha.gov/SLTC/etools/anthrax/hasp.html

References

Anthrax Clean-up in senator's office, BBC News; last visited 2/26/03
http://news.bbc.co.uk/1/hi/world/americas/1686766.stm

Figure and Table References

Figure 1.   The Search for Anthrax
http://www.fbi.gov/anthrax/searchantpicts.htm

Mold Fights Back: A Challenge for Sanitation

Ethel D. Stanley and Eli Meir

Video IX: Microbial Control

Can we manage microbial growth? Using an actual incident involving a broken water main and rapid mold growth in a new hospital as the basis for the model, this *EcoBeaker* simulation offers an opportunity to set up and then test control methods. The "Hospital Floor" model provides a virtual space to explore microbial growth patterns, set up control measures and test strategies for keeping mold at safe levels. With the "Treated Hospital Floor" model, the emergence of resistant mold raises new challenges.

Figure 1. Extensive mold growth on tiles. Uncontrolled mold growth is not only unsightly but can be harmful to human health.

Background

Microbiologists have put a lot of energy into thinking about how microbial populations grow, and there's a good reason for this—we want control! Whether we are trying to prevent the spread of an infection, preserve food, or keep our bathroom shower curtain from mildewing, we want control. Exploring how active populations respond to environmental factors and observing growth interactions is an essential part of understanding how to live with, and possibly, manage the microbes that impact our world. While we will focus on restricting growth in the hospital floor problem below, also consider that analysis of microbial reproduction models could help us understand how to enhance microbial growth.

Knowing how to support a rapidly growing microbial population can be advantageous. For example, bioremediation depends initially on our ability to nurture specific populations of microorganisms capable of removing pollutants or contaminants through biodegradation. Understanding population growth enables us to deliver sufficient numbers of genetically engineered bacteria to clean up an oil spill. An even more ambitious approach is the design of biotrickling filters that involves fixing a select community of microbes on the surface of a synthetic material. These portable "designer" biofilms metabolize a variety of undesirable organics into carbon dioxide and water in a process referred to as biofiltration.

- Describe at least three examples of desirable microbial populations and explain why each is valued.

Methods of microbial control preceded the invention of the microscope and microbiology. Microbial control is familiar to all of us, not just scientists. You can probably list a dozen or more products used in your home to control microorganisms by either killing the microbe (cidal agents) or interfering with its normal growth (static agents). Choosing an appropriate control agent and methodology is not simple.

For example, most detergents contain quaternary ammonium compounds that kill microbes by breaking the integrity of cell membranes. Detergents act as disinfectants when used to wash dishes or clean clothes, but can also act as external antiseptics when used to kill microbes on the skin. Rinsing after using detergents is required since the phosphates contained in most detergent residues may act as a nutrient source to microbes. Indeed, detergents used to clean floors and other surfaces in industrial settings have contributed to microbial growth problems. (See an example at http://www.cimcool.com/newrancd.pdf). When water containing detergent is allowed to stand or runoff on nearby floors, rancid solutions may result from actively reproducing bacteria and fungi. Clogged pipes, noxious odors and unsightly growths may produce hazardous working conditions. In addition the growth can interfere with industrial functions such as preventing metal working solutions from promoting rust resistance.

Chemical agents such as hypochlorite in solution can be effective disinfectants and in stronger concentrations can be used to sterilize surfaces. Sterilization techniques to kill and remove all microbes from an area require more drastic methods such as incineration, autoclaving, highly toxic chemicals or radiation.

Often control is best achieved by biostatic approaches that slow the growth of the microbes that are likely to be present. Interestingly, sugar is an effective static agent. As a food preservative in jams and jellies, sugar inhibits microbial growth.

- Optional pre-lab activity: Generate a chart on the blackboard listing several food preservatives and what foods you are likely to find them in. In small groups of three or four, students should choose one preservative to investigate and be prepared to explain its mode of action. If possible, describe how long the preservative has been used and where it is likely to have originated.

The Problem

"In January 1990, just prior to the scheduled opening of the Arthur G. James Cancer Hospital and Research Institute on the campus of Ohio State University, a major water pipe froze and ruptured at the roof level of the building. All twelve floors of the completely furnished building were flooded with an estimated 500,000 gallons of water. The water flowed down stairwells, elevator shafts, utility service shafts and spread out over and under each floor. Water moving over the floors wicked up into the wallboard and insulation and soaked the carpeted areas in offices, patient rooms and hallways. The water running on the undersurface of floors dropped onto the acoustical ceiling tile below. In some areas the weight of the water broke the acoustical tile insets and the water fell onto upholstered furnishings and equipment below.

Ceilings, walls, carpeted floors and upholstered furniture were either wet or exposed to high humidity due to the moisture in the building throughout the days following the flood. Removal of water and drying of surfaces was an immediate priority. It was also recognized that a conventional approach with water removal, drying, cleaning and repair would not restore the microbial integrity of the facility. To properly restore the building for its intended use as a cancer treatment facility we had to accomplish two basic objectives. Number one–eliminate the natural microbial reservoirs in building materials that had been activated by the wetting, and two–control the proliferation of fungi during demolition and reconstruction."

(See http://www.microbeshield.com/microbeshield_022.htm)

- Make a list of concerns you would want restorers to address before opening the hospital to patients, staff, and the public.

Before we get out a mop and start scrubbing or spraying a disinfectant like chlorine bleach in water, let's try and understand how fast the mold is growing in the hospital and how to control it. We'll do this by making models of how the population is growing and then try different methods of controlling the mold to see which method will work the best.

The Lab

1) Start *EcoBeaker* (click twice quickly on its icon). An installer for a demonstration version of *EcoBeaker* is located in the Mold Fights Back folder on the *Microbes Count!* web site.

When we are trying to model a serious ecological problem like this one, the best way to start is to think of the simplest model we possibly can that might explain what's going on. So we'll ignore the other surfaces, reduce the traffic on the hospital floor, and delay any consideration of the presence of other microbes. Instead, lets say that only two people walk through this part of the hospital once an hour, all day and all night. Furthermore, we'll say that each time only a couple of steps in random places on the floor will leave behind some new mold. Finally, let's assume the hospital is square and ignore all the other surfaces.

2) Select Hospital Floor from the Labs menu.

You should see a panel in the upper left of the screen labeled Hospital Floor. Right now it's empty–this is before the water main broke, and there's no mold. The floor is 25 tiles wide by 25 tiles long, for a total of 625 tiles on the floor To the right of the floor is a graph that will show how many patches of mold there are over time. At the bottom left of the window is the Control Panel, which you will use to run the models (Figure 2).

Figure 2. The Control Panel for EcoBeaker.

3) Run the model by clicking on the Go button in the Control Panel (the first green arrow). The hospital floor starts filling up with mold and the graph shows how the population is growing.

Figure 3. Simulation in progress with dark squares representing mold patches.

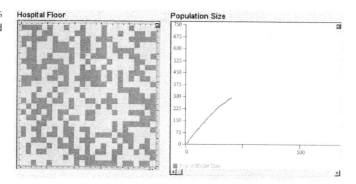

4) When the hospital floor is about half-filled with mold (reaches 300 moldy tiles), click on the red Stop button, the third icon in the control panel. Look at the graph and make a prediction about how fast the mold is going to cover the hospital floor.

- Write down how many hours you think it will take the mold to cover the floor.

5) Start the model again and when the hospital floor is filled with mold, stop the model. Look at the graph of population over time. Was your prediction of how long it would take to fill the hospital floor with mold correct? If you didn't get this right, why do you think your prediction was different than what happened?

- Explain in a few sentences why you predicted more (or less) time.

As you're looking at the graph, notice the shape of the line showing growth of the mold. The growth line starts out as almost a straight diagonal, then curves towards horizontal, and ends up being flat horizontal.

- Why does the graph have this shape?

This actually illustrates a pretty important concept in ecology. The place where the population tops out, in this case at a population of 625 mold patches, is called the **Carrying Capacity**. A population reaches its carrying capacity when it runs out of the **Limiting Resource** that it needs to keep growing.

- Common examples of limiting resources are not enough food, not enough light, and not enough nutrients. What is the limiting resource in this example?

6) Go to the Setup command on the menubar and select Model. A window showing a large folder image with tabs for Species, Habitat, Graph, Sampling, and Param (parameters) will appear. Double click on Moldy to see the smaller folder image with tabs labeled Settle, Action, and Transition Matrix.

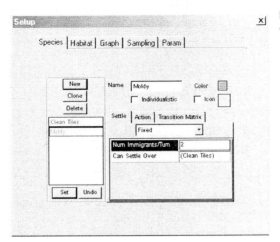

Figure 4. Species Window with Settle values for Moldy.

Notice the label Settle on the left tab of the small folders in the center of the setup box (see Figure 4). The Num Immigrants/Turn is currently set to 2, the average number of new mold colonies due to people walking through this part of the hospital every hour. (This is low, but remember, the hospital isn't open yet...)

6) To kill all the mold, you will have to spray all the floor with full strength disinfectants repeatedly at great expense. But lets say if no more than 10% of the floor has mold patches, there is no hazard to the public, patients, or staff. So the question is, how much of the mold do we have to kill each hour so that no more than 10% of the floor is covered? To figure this out, we will again simplify. We will use enough disinfectant to kill a certain percentage of the mold growing there. We will model this by using something called a Transition Matrix, which is going to specify the chance per hour that each patch of mold will die. To get to the transition matrix and add in death, click on the Transition Matrix tab.

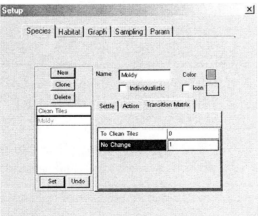

Figure 5. Transition Matrix values for Moldy Species.

There are two numbers below this, one of which is labeled Clean Tiles and the other labeled No Change. Currently there is a zero next to Clean Tiles and a one next to No Change. This means that every hour, for each moldy tile in the hospital, there is a probability of zero that the mold will die (and thus the tile will revert to being clean), and a probability of one that the mold will survive (the mold are immortal).

To simulate spraying disinfectant on some percentage of the patches, we are going to put in a probability that a patch of mold will die in each time step. Let's try killing 1% of the patches every hour. This means that the probability of a moldy tile becoming clean (the mold is killed) is 0.01, and the probability that a moldy tile will stay moldy (the mold lives - no change) is 0.99. Type these numbers into the transition matrix. Then click on Set on the lower right (Figure 6).

Figure 6. Changing the Transition Matrix values for Moldy.

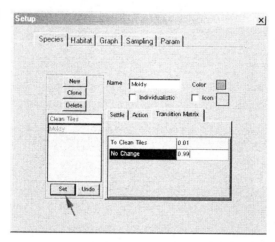

7) Before you run this model, predict what will happen. Will the mold again fill up the hospital? If not, how big will the population of mold grow?

> • Write down your prediction and a short explanation of the reasoning you used.

8) Reset the model (click on the Reset button, the fourth icon in the Control Panel) so that the hospital floor is wiped clean. Then run the model again. You will see the hospital filling up with mold again, but you'll also see some of the mold blinking out as it is killed off. Observe the graph of population size, and when it seems to have stabilized, stop the model.

> • What happened when you added in death? Does the population size stay about the same, or does it change? If it changed, did it change more or less than you expected?
>
> • What percentage of the hospital floor is covered with mold once the model has stabilized?

Our goal was no more than 10% of the floor covered by mold. If you haven't reached this goal, change the percentage of mold you are killing every hour until you get approximately 10% of the floor covered.

In order to model mold growth more realistically, we need to consider how mold spreads. The easiest extension we can make to the model is to keep two mold patches coming in every hour from the shoes, but now add a number of new mold patches every hour which depends on **Density Dependent** settlement. The number of new patches depends on how many patches are already there.

9) To change from our fixed settlement per hour to density dependent settlement we need to change the model. To do this, we need to go back into the Setup window for the mold and double click on the name Moldy. Choose Density Dependent using the scroll bar on the left side of the Settle folder.

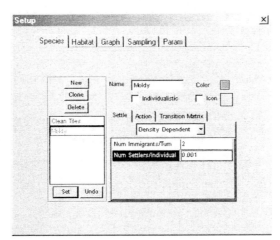

Figure 7. Changing the Settle values for Moldy Tiles.

When Density Dependent is highlighted, then let go of the mouse button. You will see that Fixed has been replaced by Density Dependent.

10) Now set the parameters of the density dependent settlement. To do this you have to enter two numbers which determine how many new mold spores settle each turn. Type in 2 for Num Immigrants / Turn, and 0.001 for the Intrinsic Rate of Growth, then leave the settlement parameters setup box (click on OK), but don't leave the larger moldy tiles setup box quite yet.

The number on the lower right is how many new mold patches each old mold patch produces per hour. Let's assume that each mold patch makes 1/1000th of a new mold patch per hour. This number 0.001 (= 1/1000) is called the **Intrinsic Rate of Growth** of the mold.

It is a good rule of thumb whenever you make a new model to start out by making it as simple as possible, so take death back out of our model. This will let us see the effect of changing just the type of growth.

11) Stop killing the mold in the model by setting the Transition Matrix back to 0 for Clean Tiles and 1 for No Change. Then click on Set.

- Now make a prediction of what the population growth over time will look like. Draw the curve for population on a piece of paper.

- What do you think the maximum population size will be? How long it will take to get there.

12) Reset the model (Reset) and then run it (Go). Watch what happens. You can save the graph results or print them out to answer the following questions. How good are you getting at predicting things? What's different? Look at the shape of the curve. What's different about that? Can you explain why we get this new shape to the curve?

- Write down your explanation.

14) Now add disinfectants back into the model to see how the treatment works with the new method of growth. Go back into the setup box for moldy tiles and add in the death rate you used before to keep the population under 10% growth. Does it still work? Does it keep the population below 10% coverage? If not, change the death rate until you manage to keep the mold below 10%.

- Have you successfully controlled the mold?

- What's your advice to the hospital staff in charge of building safety? How should they try to control the mold? If you're doing this in a class, take a couple minutes and compare your advice with that of the other groups in the class.

Notes and Comments

The principles used in these ecological models of a hospital floor are the same for mold that is spoiling crops. In that case, farmers must decide how much pesticide to use, and they must balance money versus amount of damage to the crop. Some farmers are also trying to use biological control agents to control pests such as predatory insects or bacteria and viruses that infect pests. Each of these methods has its own set of problems, and more sophisticated versions of the models you investigated here (along with lots of experiments in the real world) can help to sort out these problems. The way things grow is also fundamentally important when trying to understand natural ecosystems, and the concepts of density dependent growth, intrinsic rate of growth and carrying capacities are significant.

Lab 2: When Mold Fights Back: The Development of Resistance

In this scenario, we are revisiting the hospital after several years of using the same chemical to control mold growth. In current mold management programs, fungicide rotation is emphasized to help prevent the development of resistance. The reproduction of many generations of mold results in mutations among the progeny. By chance, a random mutation may lead to resistance to the chemical used to treat the mold.

1) Select the Treated Hospital Floor from the Markers Menu.

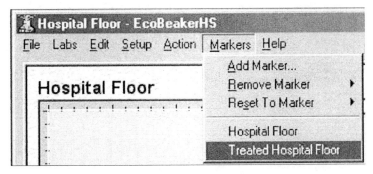

Figure 8. Markers Menu choices.

2) To see the resistant mold in this model revision, click on Setup in the menubar and then Model to open the Setup window for Treated Hospital Floor. Note the new species named Resistant Mold.

3) Note the modifications for Moldy. The Action Mutate was selected from the menu. The model displays 0.0001 for Mutate Chance. Resistant Mold is shown under Mutate. There is a 1 in 10,000 chance that Moldy will mutate into Resistant Mold each turn.

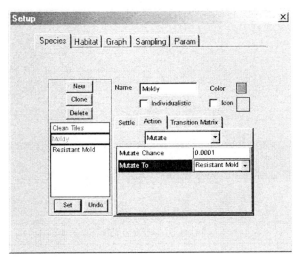

Figure 9. The Action folder in the Species Setup window for Moldy.

4) Run the simulation.

Once the resistant mold appears, the number of moldy tiles on the hospital floor quickly rises above our established limit of 10%.

Figure 10. The controlled mold (medium gray squares) and the new resistant mold (dark squares) are easily distinguished in the simulation.

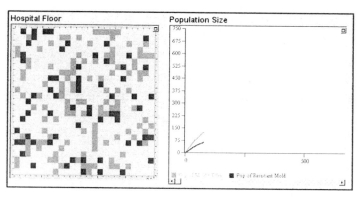

- How would you approach controlling the new resistant mold?

5) Design a Treated Hospital Floor model that uses this approach. Run the simulation.

- Were your results as expected?

Software Used in this Activity

EcoBeaker HS Demo
   Eli Meir (University of Washington)

Platform Compatibility: Macintosh and Windows

Additional Resources

Available on the *Microbes Count!* web site at http://bioquest.org/microbescount

Software
EcoBeaker HS Demo

Text
A PDF copy of this activity, formatted for printing

Related *Microbes Count!* Activities

Chapter 3. Modeling Wine Fermentation

Chapter 3. Modeling Microbial Growth

Chapter 4. TB and Antibiotic Resistance

Chapter 9. Bioterror and the Microbiologist's Response

Chapter 10. Controlling Potato Blight: Past, Present, and Future

Unseen Life on Earth Telecourse

Coordinates with Video IX: Microbial Control

Relevant Textbook Keywords

Carrying capacity, Cidal agent, Disinfectant, Limiting resource, Mutation, Resistance, Sanitation, Static agent,

Related Web Sites (accessed on 4/29/03)

Aeromicrobial Control in an Extensively Damaged Hospital
http://www.microbeshield.com/microbeshield_022.htm

American Society for Microbiology
http://asmusa.org

Antimicrobial Resistance, NIAID Fact Sheet
http://www.niaid.nih.gov/factsheets/antimicro.htm

Antimicrobial Resistance: NIH Response to a Growing Problem
http://www.niaid.nih.gov/director/congress/1999/0225.htm

Centers for Disease Control and Prevention: Promoting Appropriate Antibiotic Use in the Community
http://www.cdc.gov/drugresistance/community

Control of Rancidity in Metal Working Fluids
http://www.cimcool.com/newrancd.pdf

Controlling the growth of microorganisms
http://www.bact.wisc.edu/microtextbook/ControlGrowth/chemAgent.html

Microbes Count! Website
http://bioquest.org/microbescount

Sick Building Syndrome
http://gtresearchnews.gatech.edu/reshor/rh-sf95/fungus.htm

Unseen Life on Earth: A Telecourse
http://www.microbeworld.org/htm/mam/is_telecourse.htm

VOC Control: Biotrickling filters
http://www.prdtechinc.com/voc.html

References

Meir, Eli (2003). *EcoBeaker HS Demo.* SimBiotic Software. Ithaca, NY
http://www.ecobeaker.com/

Figure and Table References

Figure 1.   Courtesy of HKC and Associates Inc.
http://www.hkcenv.com/mold.htm

Citrus Canker: Alternatives for Control

Linda Weinland, Peter Woodruff, Margaret Waterman, and Ethel Stanley

In the following case, Family Trees, you are asked to consider not only the relevant scientific issues involved in the control and eradication of citrus canker, but also the social, economic, and human implications of microbial control efforts.

Begin by asking one member of your group to read the case out loud. Use the case analysis worksheet on the next page to work through the case. (For more information on using cases, read the introduction to "The Farmer and the Gene" in Chapter 5.)

- Could your group work with citrus canker in the laboratory? Explain.

Family Trees

Carlos Silva sipped his morning coffee in the shade of the orange and grapefruit trees in the yard. He had planted one at the birth of each of his grandchildren and enjoyed seeing how much each had grown over the past eight years. Except for a few broken branches on the trees farthest from his house, all had survived another hurricane season.

Glancing down at the paper, Carlos was startled by the full-color map on the front page. His eyes moved quickly to the center of the map. With a sigh of relief, he found that his own home was clearly outside a yellow zone north of US 41 and east of NW 87th Ave.

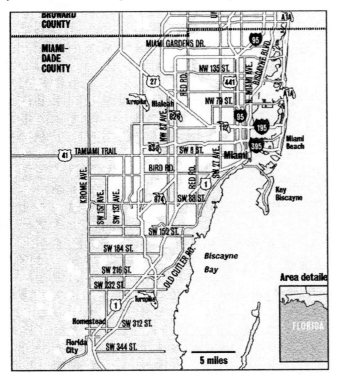

Figure 1. Citrus Canker Zones

- Lighter areas are 1900 ft canker zones.

- If your citrus tree is located in one of the lighter map areas it will likely be cut down by the state citrus-canker fighters.

http://www.citruscanker.com/canker0726map.gif

Miami Herald, July 26, 2001

Peeling the grapefruit he had just picked this morning, Carlos wondered if the small brown spots on the rind were a problem.

Figure 2. Label from a crate of clementines purchased at a Philadelphia fruit market. Of what significance is the phrase, *Not for distribution in AZ, CA, FL, LA, TX, Puerto Rico, and any other U.S. Territories?*

Worksheet for the Citrus Canker Case Analysis

(A printable copy of this worksheet is available on the *Microbes Count!* web site.)

1. Recognize potential issues. List terms or phrases that seem to be important for understanding what the case is about.

2. Brainstorm for connections. Briefly discuss the following with the group.

    What is this case about?

    What are its major themes?

    Keep track of major issues and questions that arise with the Know/Need To Know chart.

    Identify one question or issue from the "need to know" list that your group wants to explore.

What do we already know?	What do we still need to know?

3. Obtain additional references or resources to help answer or explore questions. These may include print resources, informational articles, data sets, results of simulations, maps, interviews, etc.

    List four different resources you think would be important to use to learn more.

4. Design and conduct scientific investigations relevant to the question.

    Investigations could be laboratory, field or computer-based experiences that the instructor arranges for the entire class or may be entirely student generated. Describe your plans.

The BioQUEST Curriculum Consortium   Linda Weinland, Peter Woodruff, Margaret Waterman, and Ethel D. Stanley

5. Communicate your understanding of the outcomes of your work on the question.

    Produce materials which outline the problem and support your conclusions. These artifacts can take many forms, from traditional papers and scientific reports, to posters, videos, pamphlets, consulting reports, role playing, interviews, etc. Decide what products would be best for this case investigation.

Figure 3. A collection of citrus products from the local supermarket. A quick scan of the labels reveals the origin of these citrus products to include Florida, California, Texas, Mexico, Brazil, Argentina, Curacao, Portugal and China.

Citrus Canker Case Team Projects

The following were identified as research questions of interest by students in environmental biology at a local college.

1. What is citrus canker?

    Describe the biology and life cycle of the causative agent(s).

    What are the possible host species?

    Does the canker organism (or its by-products) affect humans?

2. When and where was citrus canker first discovered in the world? In the U.S.?

    Where in the world and in the U.S. does it occur today? (Include maps.)

    How does it spread?

3. Describe all aspects of historical and current control measures in the U.S.

    Have they been effective?

4. Describe all aspects of historical and current control measures in Brazil.

    Have they been effective?

5. Describe the economic impact of the citrus industry in Florida, the U.S., and at least two other countries in the world.

6. How can biotechnology assist scientists attempting to solve the citrus canker problem?

Linda Weinland, Peter Woodruff, Margaret Waterman, and Ethel D. Stanley   The BioQUEST Curriculum Consortium

Figure 4. This model looks at five different alternatives proposed to control citrus canker with respect to political, social, economic, and ecological goals.

Optional Activity: Decision Making and Control of Citrus Canker

There are several options that could be used to deal with the problem of citrus canker. You can construct decision making models to examine these options at the Web-HIPRE Hierachical Preference Analysis site at http://www.hipre.hut.fi. A sample model is shown in Figure 4.

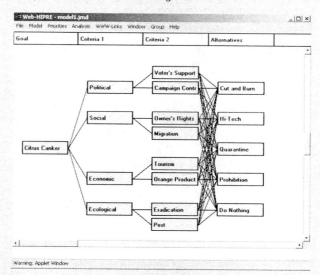

Possible options for the control of citrus canker include:

1. Cut and burn potentially infected trees as needed regardless of ownership.

2. Fund research to look for high tech solutions to control citrus canker.

3. Quarantine areas with infected trees to provide barriers to transmission.

4. Prohibit the planting of new citrus trees by homeowners

5. Do nothing

You should:

- Identify several community stakeholders in a town facing citrus canker control

- Be prepared to role play one of these stakeholders

- Discuss the control methods and the impact each has on the goals from the stakeholder's perspective.

Web Resources in this Activity

Web-HIPRE Hierachical Preference Analysis site
http://www.hipre.hut.fi

Additional Resources

Available on the *Microbes Count!* web site at http://bioquest.org/microbescount

Text

A PDF copy of this activity, formatted for printing

Worksheet for the Citrus Canker Case Analysis

Related *Microbes Count!* Activities

Chapter 5: The Farmer and the Gene: A Case Approach to Bt Corn

Chapter 12: Souvenirs: Investigating a Disease Outbreak

Unseen Life on Earth Telecourse

Coordinates with Video IX: Microbial Control

Relevant Textbook Keywords

Agriculture, Control, Pathogen

Related Web Sites

Citrus Leaf Miner Life Cycle
http://primera.tamu.edu/citrus_entomology.htm
http://primera.tamu.edu/citrus_leafminer_parasites.htm

Citrus Canker Maps
http://www.citruscanker.com/canker0726map.gif
http://doacs.state.fl.us/canker/cankerflorida.pdf
http://doacs.state.fl.us/canker/images/canker-by-year.jpg
http://www.nass.usda.gov/fl/gif/cprd0001.gif

Control of Citrus Canker
http://www.imok.ufl.edu/plant/docs/canker_indian_river_2002.pdf
http://cancer.lbi.ic.unicamp.br/xanthomonas/citri/Maps/M2/1.1.html
http://www.agriculture.com/default.sph/agNotebook.class?FNC=ArticleList__Aarticle_html___3097___7
(note: there are multiple underlines in this URL)
http://preserve.nal.usda.gov:8300/jag/v14/v14i9/140337/a140337.htm

Hurricanes
http://www.escambia-emergency.com/images/hcanemap.jpg
http://www.fema.gov/nwz99/images/hurr0819.gif
http://www.met.fsu.edu/explores/Tropical/atl2.htm l
http://www.edrinc.com/edr-hurricane.jpg

Xanthomonas axonopodis pv. *citri* Gene Map
http://cancer.lbi.ic.unicamp.br/xanthomonas/citri/Maps/1.1.html

Linda Weinland, Peter Woodruff, Margaret Waterman, and Ethel D. Stanley   The BioQUEST Curriculum Consortium

Figure and Table References

Figure 1.   Modified from the *Miami Herald, July 26, 2001*
http://www.citruscanker.com/canker0726map.gif

Figure 2.   Courtesy of Ethel Stanley

Figure 3.   Courtesy Margaret Waterman

Figure 4.   Screen shot of citrus canker scenario using Web-HIPRE
Hierachical Preference Analysis
http://www.hipre.hut.fi

A Plague on Both Houses: Modeling Viral Infection to Control a Pest Outbreak

Anton E. Weisstein

Video IX: Microbial Control

Background on myxoma

Myxoma is a virus that occurs naturally in the South American jungle rabbit *Sylvilagus brasiliensis*. Infection with myxoma virus leads to the disease myxomatosis; this disease is mild in *S. brasiliensis* but almost always lethal in the European rabbit *Oryctolagus cuniculus* (Fenner and Ratcliffe, 1965).

The European rabbit was imported to Australia in 1859 and rapidly became a serious agricultural pest. In an attempt to control the rabbit population, myxoma virus was introduced in Australia in 1950. Mosquitoes quickly spread the disease, decimating rabbit populations: over 99% of rabbits that became infected with myxoma died of the infection (Fenner and Ratcliffe, 1965).

Figure 1. Devastation of pastureland by rabbits. The field on the right is protected by a rabbit-proof fence.

Within a year, however, the mortality rate had dropped to 90%, and it continued to decrease steadily for over a decade. Laboratory studies showed that two factors were responsible: rabbits were evolving resistance to the virus (Marshall and Douglas, 1961), and the virus was simultaneously becoming less deadly (Fenner and Ratcliffe, 1965).

In the end, strains of the virus with mortality rates between 70-95% came to predominate over both milder and more virulent strains (Fenner and Ratcliffe, 1965), leading to partial control of the rabbit population. Further control measures have been undertaken, including introduction of new myxoma vectors (the European and Spanish rabbit fleas), release of a second biological agent (rabbit calcivirus), and physical control methods such as poisoning and shooting. The result is that rabbit numbers are roughly half of what they were in the late 1940s (Williams *et al.*, 1995). Despite all these efforts, however, rabbits continue to cause approximately $600 million (Australian) in agricultural damages each year (Department of Agriculture —Western Australia 2001).

Background on epidemiological models

In 1979, Anderson and May presented a theoretical epidemiological construct which has since become known as an SIR model. The SIR model forms the basis for the *Epidemiology* computer simulation, which is used in several other activities in this book. (See the list of related *Microbes Count!* activities at the end of this lab.) Although the questions posed by the myxoma model could be answered using the simulations in *Epidemiology*, in this activity you will have the opportunity to explore the mathematical model that underlies the computer generated simulations and to develop a deeper understanding of the SIR model.

Under the SIR model, the host population is partitioned into three categories: susceptible individuals (*S*), infected individuals (*I*), and individuals who have recovered from infection and thereby become immune to re-infection (*R*). This framework enables us to describe and predict the course of an epidemic by tracking movement into, out of, and between these compartments. For example, an individual's recovery from infection can be represented as decreasing the value of *I* by one and increasing the value of *R* by one. Similarly, if a susceptible individual is born, we note this event by increasing *S* by one.

Among the most important parameters of such a model are:

• The natural birth and death rates of the host,

• The rates at which hosts die and recover from infection, and

• The rate at which infection spreads from infected to uninfected hosts.

Depending on the model details, additional parameters may also be necessary. In studying the myxoma case, for example, our model should include multiple strains of the virus, each of which may have different host mortality, recovery, and transmission rates. It is also important to consider the manner in which a particular disease is spread. Infections easily spread through casual contact, such as flu and smallpox, spread at rates proportional to the numbers of both susceptible and infected host individuals ($= S \cdot I$). Other infections, such as malaria and myxoma, are transmitted by vectors rather than from host to host; the spread of these diseases therefore depends not only on the numbers of susceptible and infected hosts (*S* and *I*), but also on the numbers of uninfected and infected vectors (*U* and *V*). Myxoma virus can be transmitted by several vectors including mosquitoes, fleas, and blackflies; for the sake of simplicity, however, our model will include only a single vector species.

Myxoma model

For the sake of simplicity, we will use a discrete-time form of the SIR model. Each time interval therefore reflects a set period of objective time such as a day or a week. We will also make the following simplifying assumptions, not all of which may be biologically realistic:

• Asexual reproduction in both hosts and vectors.

• No population age structure.

• At most one strain can infect a single host or vector individual.

• Recovery from *any* strain confers permanent immunity to *all* strains.

• Infection and recovery do not affect reproductive rates.

• Neonatal infection in hosts but not in vectors.

• No inherited immunity.

• Vectors not affected by infection.

• Infectious vectors remain permanently infectious.

• Positive correlation between virulence and transmissibility (see below).

Again for the sake of simplicity, our model will include only three strains of the virus. The host population will therefore be divided into five compartments: susceptible (S), infected with strain #1, 2, or 3 (I_1, I_2, I_3), and recovered (R). Similarly, the vector population will be divided into four compartments, one for uninfected vectors (U) and one for vectors infected with each of the three strains (V_1, V_2, V_3). Figure 2 shows how individuals move between these compartments.

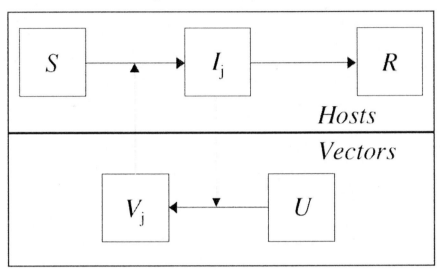

Figure 2: Compartments of the host and vector populations. Hosts move from susceptible to infected to recovered; vectors from uninfected to infected. The dotted lines across populations denote the cycle of virus transmission; thus, transmission by infected vectors causes susceptible hosts to become infected, and transmission by infected hosts causes uninfected vectors to become infected.

We assume logistic growth of the host population. The easiest way to do this is to make the host birth rate *decrease* linearly with population size and the host death rate *increase* linearly with population size. (See Figure 3.)

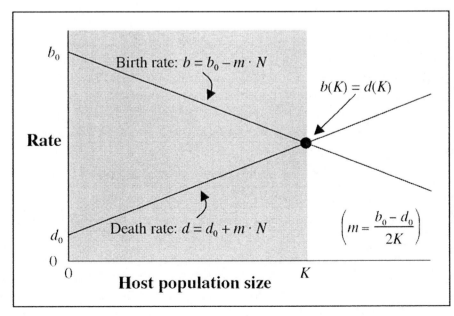

Figure 3: Birth and death rates for a population undergoing logistic growth. As the population size increases, the birth rate falls and the death rate rises. The two lines representing these rates intersect at the population's equilibrium, namely the carrying capacity (K). Below carrying capacity, the population's birth rate exceeds its death rate, so the population will increase; this area is shaded grey. Above carrying capacity, the inequality is reversed, so the population will decrease; this area is shaded white.

Table 1: Key parameters of the model.

Name	Parameter	Representative value	Units
b_0	Host birth rate at N = 0	0.04	births / host
d_0	Host (natural) death rate at $N = 0$	0.01	deaths / host
K	Host carrying capacity	10,000	host individuals
k_j	Host death rate from infection with strain j	0.4	deaths / infected host
r_j	Host recovery rate from infection with strain j	0.2	recoveries / infected host
b'	Vector birth rate	0.2	births / vector
d'	Vector death rate	0.2	deaths / vector
ω	Host-vector contact rate	0.00003	contacts / host / vector
α_j	Vector → host transmission rate of strain j	0.4	transmissions / infected vector / susceptible host
β_j	Host → vector transmission rate of strain j	0.9	transmissions / infected host / susceptible vector

Table 1 lists key parameters of the model. It's important to understand the interpretation of each parameter so that you can choose biologically meaningful values when running your own simulations. In particular, the host-vector contact rate ω *must* be very small or the simulation will give meaningless results, such as population sizes less than zero. Each strain must also satisfy

$$k_j + r_j \le 1 - d$$

which states that an infected individual can't die twice, and can't both die and recover.

Finally, some model parameters may be interdependent. For example, many infectious diseases produce a variable viral titer in the host: cases with high titer are generally both more lethal and more easily transmitted to a vector. Myxomatosis follows this general pattern (Fenner and Ratcliffe 1965), so we should choose values k_j and β_j that are positively correlated across strains. It is convenient to order the strains from most virulent (#1) to least virulent (#3); then we need only ensure that $k_1 > k_2 > k_3$ and $\beta_1 > \beta_2 > \beta_3$.

We can describe the model outlined above by the following five recursion equations (Table 2). Each of these equations gives the number of individuals in one compartment (e.g., susceptible hosts) in terms of the model parameters and the compartment sizes at the previous time interval. I urge you to take a few minutes to look at these equations in the context of Figure 2 and see how they encapsulate the model's assumptions. You may find it helpful to recall that addition combines *independent* terms (e.g., newborns vs. adults), multiplication combines

Table 2. Recursion equations used by the model.

$$
\begin{array}{ccccc}
\overbrace{\text{New number}}^{} & \overbrace{\text{Susceptible}}^{} & & \overbrace{\text{Existing}}^{} & \overbrace{\text{and don't}}^{} \\
\text{of susceptible} = & \text{newborn} & + & \text{susceptible} \quad \text{who} & \text{become} \\
\text{hosts} & \text{hosts} & & \text{hosts...} \quad \text{don't die...} & \text{infected.}
\end{array}
$$

$$(1) \quad \overbrace{S(t+1)}^{} = \overbrace{b[S(t)+R(t)]}^{} + \overbrace{S(t) \cdot (1-d)}^{} \cdot \overbrace{\left[1-\omega \sum_j \alpha_j V_j(t)\right]}^{}.$$

$$
\begin{array}{ccccccc}
\text{New number of} & \text{Infected} & \text{Existing} & \text{who don't} & \text{Existing} & \text{who} & \text{but do} \\
\text{hosts infected} = & \text{newborn} + & \text{infected} & \text{die or} & + \text{susceptible} & \text{don't} & \text{become} \\
\text{with strain } j & \text{hosts} & \text{hosts...} & \text{recover} & \text{hosts...} & \text{die...} & \text{infected.}
\end{array}
$$

$$(2) \quad I_j(t+1) = b \cdot I_j(t) + I_j(t)\cdot(1-d-k_j-r_j) + S(t)\cdot(1-d)\cdot[\omega\, \alpha_j V_j(t)].$$

$$
\begin{array}{cccc}
\text{New number} & \text{Existing} & \text{who} & \text{Number of hosts} \\
\text{of recovered} = & \text{recovered} & \text{don't} & + \text{recovering from} \\
\text{hosts} & \text{hosts...} & \text{die} & \text{any strain.}
\end{array}
$$

$$(3) \quad R(t+1) = R(t) \cdot (1-d) + \sum_j r_j I_j(t).$$

$$
\begin{array}{ccccc}
\text{New number} & \text{Newly} & & \text{Existing} & \text{and don't} \\
\text{of uninfected} = & \text{born} & + & \text{uninfected} \quad \text{who} & \text{become} \\
\text{vectors} & \text{vectors} & & \text{vectors...} \quad \text{don't die...} & \text{infected.}
\end{array}
$$

$$(4) \quad U(t+1) = b'\left[U(t)+\sum_j V_j(t)\right] + U(t) \cdot (1-d') \cdot \left[1-\omega \sum_j \beta_j I_j(t)\right].$$

$$
\begin{array}{ccccc}
\text{New number of} & \text{Existing} & \text{who} & \text{Existing} \quad \text{who} & \text{but do} \\
\text{vectors infected} = & \text{infected} & \text{don't} + & \text{uninfected} \quad \text{don't} & \text{become} \\
\text{with strain } j & \text{vectors...} & \text{die} & \text{vectors...} \quad \text{die...} & \text{infected.}
\end{array}
$$

$$(5) \quad V_j(t+1) = V_j(t) \cdot (1-d') + U(t) \cdot (1-d') \cdot [\omega \beta_j I_j(t)].$$

conditional terms (e.g., susceptible hosts who become infected), and subtraction from one represents logical *negation* (e.g., one minus the proportion dying represents the proportion surviving).

Understanding and using the myxoma workbook

The mathematical model developed above is contained in the Microsoft *Excel*® workbook "Myxoma template.xls", located in the Myxoma section of the *Microbes Count!* web site. All the equations have already been entered into the spreadsheet, so all you need to do is choose the parameter values you want, run the model, and interpret the results. See Figure 4.

Columns A-R of row 3 show the state of the host and vector populations at time zero. Scroll to the right to view the initial number of susceptible hosts, hosts infected with each strain, recovered hosts, total hosts, % hosts infected with each strain, number of uninfected vectors, and number of vectors infected with each strain. Some cells are outlined in red: these are initial conditions that you can

Figure 4: Screen shot of the myxoma workbook, including partial data table and graphs. See text for details.

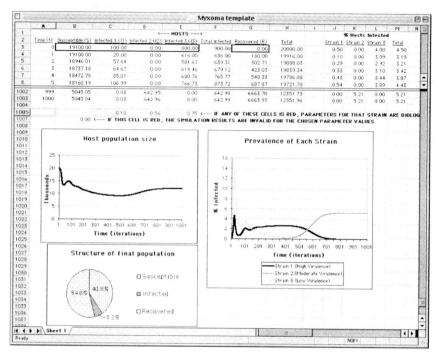

modify to perform your own simulation. Values in other cells, such as F3 ("Total infected"), are computed from the initial conditions and should not be entered directly.

Column U contains the model parameters such as birth and death rates, carrying capacity, and virulence/transmissibility parameters for each strain. Again, these cells are outlined in red to indicate that new values can be entered into these cells. Note that cell U7, the host-vector contact rate ω, is shaded red to remind you that errors will result unless this value is kept very small, especially when host and/or vector numbers are large.

Rows 4 through 1003 contain iterations of model equations (1)-(5). The screen has been split vertically so you can see both the first few and the last few iterations. The bottom half of the screen contains four "warning" cells that will turn red in case of simulation errors. Be sure to check these cells before interpreting any results!

Also shown are three graphs. "Host population size" plots the size of the host population over time. "Prevalence of each strain" plots the proportion of the host population infected with each of the three strains over time. "Structure of final population" shows the proportion of the host population at time t = 1000 who are susceptible, infected, and recovered. As you conduct your own investigations, you may find it useful to construct additional graphs using *Excel*'s ChartWizard function.

As soon as you change any starting condition or parameter, the workbook will immediately run a simulation using the new value and graph the results. If you want to change several values before running the simulation (for example, if you are setting the initial conditions equal to the final conditions of a previous run), you must first disable this feature by setting calculations to manual mode. The details of how to do this vary among versions of Excel: consult your version's Help Index for specific instructions. Once calculations are set to manual, the workbook will recalculate only when you use the "Calculate Now". command (see Excel's help Index) or set calculations back to automatic mode.

Investigations

1. Set the host population's carrying capacity (K) to 20,000 and introduce a small number of individuals infected with strain 1. Notice the changes in the graphs at the bottom of the worksheet. At what value does the population size stabilize? Why is this less than the carrying capacity? Does the population converge smoothly to that value? How would you interpret the fluctuations in virus prevalence over time?

   Repeat for strains 2 and 3 and compare across strains, recalling that strain #1 is most virulent and strain #3 least virulent. Which strain leads to the greatest reduction in population size? Which strain leads to the greatest % of infected individuals at equilibrium? Are these patterns what you would have expected? Discuss possible explanations.

2. Reduce K to 12,000 and repeat the above analysis for each strain. What change do you note in the % infected at equilibrium? What do you think might happen if you reduced the carrying capacity further? Try it and see.

   What you have observed is the epidemiological threshold. Contagious diseases need a certain minimum number of hosts present or else they are lost, although the specific threshold varies from one disease to another. Which of the three strains you examined had the highest threshold and which the lowest? What are the implications of epidemiological thresholds for small and isolated populations? When and how have epidemiological thresholds influenced human history? You may find recent papers by Mitchell and Power (2003) and Torchin et al. (2003) of interest in this context.

   Imagine that we are trying to control a disease by eradicating its vector. Set K back to its original value of 20,000 and reduce the number of vectors present by 20%. What effect does this change have on the number of infected hosts at equilibrium? Continue reducing the size of the vector population. By how much must you reduce the vector population to eliminate the disease? Is it the same across all three strains? If not, which strain is easiest to eliminate and which is hardest? Is this the same pattern you saw for the strains' host epidemiological threshold? Why or why not?

3. Begin with only strain 1 present and run the worksheet until the system reaches equilibrium. Note the host population size and % infected. Now model the appearance of a mutant strain of the virus by introducing a

single individual infected with strain 2. Run the worksheet, paying particular attention to the graph of % infected. Describe any changes you see. How would you interpret them? Would you obtain the same result if the virus had mutated to strain 3 instead of strain 2? To what extent do your observations match what actually happened with myxoma in Australia?

Try altering the parameters associated with each strain (k, r, α, β), retaining the positive correlation between virulence (k) and transmissibility (β). Repeat the above analysis several times for different parameter values. Does the scenario play out in the same way? What are the implications for understanding the evolution of virulence?

4. As Health Minister for the island nation of Knessy, you know that there are two strains of the mosquito-borne morbidia virus circulating among your citizens, one of which (strain X) is much more lethal than the other (strain A). Fortunately, infections with strain X are sporadic and never seem to spark a widespread epidemic. Still, even the relatively mild illness produced by strain A is a drain on your economy and a cause of significant suffering.

    You have identified four promising intervention strategies to deal with this problem:

    • Formulate an advanced treatment that will speed recovery from strain A.

    • Drain a nearby marshland that harbors a large fraction of the local mosquito population.

    • Develop and release a weakened strain of the virus.

    • Change people's behavior: limit outdoor gatherings, encourage use of insect repellent, etc.

    Unfortunately, your budget will allow you to pursue only one of these strategies. Which one do you choose? Can you suggest other strategies that might be even more successful? Justify your decision based on simulation results as well as any other considerations (e.g. political, ecological, economic) that might be relevant. What additional data might you want to collect before making your decision?

5. Discuss the model's assumptions. Are any of them unrealistic? How might you reformulate the model to incorporate more realistic assumptions? To what extent would this complicate the model's implementation and interpretation?

    What factors influence the relative competitive dominance of different strains? Is it possible to develop a formula for determining which of three strains would be competitively dominant?

    What differences would there be between the vector-based myxoma model and a model of a directly transmissible disease such as influenza? In what ways might these structural differences cause the two models to generate different predictions?

Additional Resources

Available on the *Microbes Count!* web site at http://bioquest.org/microbescount

Text

A PDF copy of this activity, formatted for printing

The "Myxoma template" model file

Related *Microbes Count!* Activities

Chapter 1: Modeling More Mold

Chapter 1: Population Explosion: Modeling Phage Growth

Chapter 6: Tracking the West Nile Virus

Chapter 8: Investigating Predator-Prey Interactions

Chapter 10: Measles in Nakivale Refugee Camp

Chapter 11: Epidemiology: Understanding Disease Spread

Chapter 12: Vaccine: Experimenting with Strategies to Control Infectious
Diseases

Unseen Life on Earth Telecourse

Coordinates with Video IX: Microbial Control

Relevant Textbook Keywords

Biological control, Epidemiology, Mathematical modeling

Related Web Sites (accessed on 4/18/03)

Economic and Ecological Impact of Rabbits in Australia
http://rubens.anu.edu.au/student.projects/rabbits/home.html

Microbes Count! Website
http://bioquest.org/microbescount

Unseen Life on Earth: A Telecourse
http://www.microbeworld.org/htm/mam/is_telecourse.htm

References

Anderson, R. M., and R. M. May (1979). Population biology of infectious
diseases. *Nature* 280: Part I: 361-367 and Part II: 455-461.

Department of Agriculture—Western Australia (2001). Farmnote 25/2001:
European wild rabbit.
http://www.agric.wa.gov.au/agency/pubns/FARMNOTE/2001/f02501.htm

Dwyer, G., S. A. Levin, and L. Buttel (1990). A simulation model of the
population dynamics and evolution of myxomatosis. *Ecological
Monographs* 60: 423-447.

Fenner, F., and R. N. Ratcliffe (1965). *Myxomatosis*. Cambridge University Press, London, England.

Marshall, I. D., and G. W. Douglas (1961). Studies in the epidemiology of infectious myxomatosis of rabbits. VIII. Further observations on changes in the innate resistance of Australian wild rabbits exposed to myxomatosis. *Journal of Hygiene* 59: 117-122.

Mitchell, C. E., and A. G. Power (2003). Release of invasive plants from fungal and viral pathogens. *Nature* 421: 625-627.

Torchin, M. E., K. D. Lafferty, A. P. Dobson, V. J. McKenzie, and A. M. Kuris (2003). Introduced species and their missing parasites. *Nature* 421: 628-630.

Williams, K., I. Parer, B. Coman, J. Burley, and M. Braysher (1995). Managing vertebrate pest: rabbits. Bureau of Resource Sciences/CSIRO Division of Wildlife and Ecology, Australian Government Publishing Service, Canberra. p. 284.

Figure and Table References

Figure 1.   Included with permission, Fenner and Ratcliffe (1965)

Figure 4.   Screen shot from the "Myxoma Template.xls" model file.

Anton E. Weisstein

Chapter 10

Activities for Video X: Microbial Interactions

We are amazed at the interactions between microbial populations in the rumen of the cow as well as the adaptive microbes that help termites digest wood meals, yet our view of microbial interactions is limited by the systems we choose to access and study. Often it is the problematic microbes such as Phytophthora ramorum, the cause of an outbreak of Sudden Oak Death Syndrome in California during 2002 that attract our attention.

In this unit we can:

- make a profit or lose the farm as you investigate the economic consequences of using chemical control approaches to managing late blight in potatoes,
- manipulate environmental variables in a life cycle model of a normally benign dinoflagellate to produce an outbreak of its toxic form, and
- explore factors that may have contributed to a measles outbreak at a refugee camp.

Controlling Potato Blight: Past, Present, and Future

Potato late blight, Phytophthora infestans, has an infamous past, yet it continues to present a challenge to modern day farmers. Historical scenarios in the LateBlight simulation help us define the impact of this disease before the interactions between this microbe and the potato were understood. Modern scenarios enable us to investigate current strategies to control this pathogen from the management of cull piles to the use of genetically engineered potatoes. A life cycle model, Potato Late Blight, provides an additional method for exploring microbial interactions.

Making Sense of the Complex Life Cycles of Toxic Pfiesteria

An intriguing microbe with complex interactions that scientists have recently begun to study is the dinoflagellate Pfiesteria piscicida. It has been linked to large fish kills in the waters of North Carolina and the Chesapeake Bay. Examine the effect of environmental variables associated with outbreaks such as flooding and fish density with a model of the Pfiesteria life cycle.

Measles in the Nakivale Refugee Camp: Epidemiology in Action

In the spring of 2000, the Nakivale Refugee Camp in Southwestern Uganda experienced a serious outbreak of measles. Factors such as movement of populations, social interactions, and low immunization levels influenced this outbreak. Model the spread of measles in the population and assess the impact of these factors with the Epidemiology program.

Controlling Potato Blight: Past, Present, and Future

Ethel D. Stanley and Philip Arneson

Video X: Microbial Interactions

The management of disease organisms to ensure productivity in crop plants has been a human endeavor throughout our agricultural history and science has greatly informed our modern methods. Currently, an epidemic of the Cassava Mosaic Virus threatens the human population in Zimbabwe by significantly reducing the production of edible cassava roots. Famine is a distinct possibility as the cassava root is a major component of the daily diet for these inhabitants. One promising control effort for Cassava Mosaic Virus is the genetic engineering of cassava plants to enhance resistance to the virus.

In the 1840's, the spread of a crop disease through Ireland resulted in mass starvation and migration of the inhabitants. This disease of potatoes is known as Late Blight and remains a serious agricultural problem today. Under favorable environmental conditions and in the absence of any control measures, it can destroy virtually 100% of the foliage of susceptible cultivars.

In the 1840's little was known about this disease. Today we know that Late Blight is caused by the microorganism *Phytophthora infestans*. The spores of *P. infestans* are easily disseminated by wind and rain from infected foliage. Further, spores can survive the winter and other unfavorable conditions in lesions on potatoes and can infect a field through:

- new "seed" tubers planted at the beginning of the season,

- "volunteer" tubers that were left after harvest in the field, and

- piles of "culled" potatoes that have been discarded.

Figure 1. A family faces the consequences of the mid-19th century potato famine in Ireland.

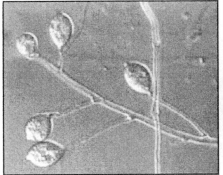 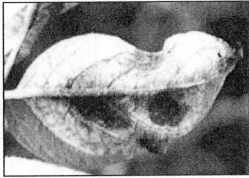

Figure 2. A micrograph showing *Phytophthora infestans* sporangia.

Figure 3. Late season leaf damage in blighted potato plant.

LateBlight Investigations

Using a simulation program called *LateBlight*, you will be asked to simulate potato farming in 19th century Ireland and answer some specific questions about your harvests. Then you will explore modern management strategies and report harvest data as a modern potato grower. (An installer and brief instructions for the LateBlight application (Microsoft Windows only) is located in the *Potato Blight* folder on the *Microbes Count!* web site.

Figure 4. Screen shot of the *LateBlight* simulation program. Daily data throughout the potato growing season are displayed for temperature (upper plot), leaf production (middle), rainfall (lower), and blight (starts in August).

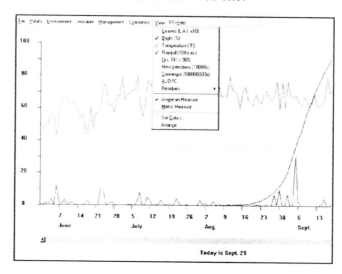

Part 1: Growing Potatoes in the 1840's

The Irish grew potatoes in Ireland for over two hundred years with good success. It wasn't until the 1840's that potato farming became far more problematic. Let's look at the results of three different seasons: 1830 with no late blight; 1843 when the first late blight occurred in the country; and 1844.

Figure 5. *LateBlight* icon.

a. Double click the Begin *LateBlight* potato icon. Go to File in the menu and choose Open. Highlight the file "past1.lbt" and the results of the 1830 season will appear. Resize the window by clicking in the middle box in the upper right or by dragging the lower right corner of the window. Choose Economics from the menu and highlight Show Report. (Note: Only the Yield information is relevant here.)

• How many potatoes in kilograms per hectare can you expect to harvest in 1830?

b. Open the file "past2.lbt" and examine the results of the 1843 season. When do you begin to see signs of late blight infection?

Choose Economics from the menu and highlight Show Report.

• Now, how many potatoes can you expect to harvest in 1843?

c. Open the file "past3.lbt" and look at the results of the 1844 season. When do you begin to see signs of late blight infection?

The agricultural methods of the day contribute to your problems. Volunteer potatoes (tubers left in the ground over winter) were common as farmers simply dug only what they needed as the season progressed. Also cull piles (damaged tubers that weren't eaten) were found near the fields since no one knew how this disease spread.

Choose Economics from the menu and highlight Show Report.

- How many potatoes can you expect to harvest in the third season?

Part 2: Potato Farming in the 1990's

Management of potato late blight begins by reducing the initial inoculum in the field by:

1. Keeping the numbers of infected seed tubers to very low levels by planting only certified seed,

2. Crop rotation to reduce the numbers of volunteer plants (which may be infected), and

3. Burying or composting cull piles or at least keeping them a great distance from the potato fields.

You are farming in up-state New York during the 1990's. Despite following the management practices described above, you have been having trouble with late blight. The county has had unusually cool, wet weather for the past three years. Encouraged by reports of a new hybrid potato that has high resistance to late blight, you are thinking of trying these potatoes yourself. The new seed certified potatoes will cost you about $100 more per acre than your standard low resistance potatoes.

a. Open the file "potato1.lbt". This scenario was created by selecting Potato and changing Resistance Level from low to high and then selecting Economics and increasing Costs from $1100 to $1200 per acre. (Note: In the United States, the yield is calculated in cwt or hundredweights per acre, i.e.. 300 cwt/acre equals 30,000 pounds/acre.) Review the Economics Report results.

- Did you make a profit? (Show amount.)

b. What if the county experienced a moderately dry growing season instead?

Open the file "potato2.lbt", which was created by using the weather file "moddry.lwx" and keeping all the same values you used above. Review the results.

- Did you make a profit? (Show amount.)

c. The uncertainty of the weather raises new issues for you. You decide to try a more vigorous spray program instead of planting the high resistance potato. Open the file "potato3.lbt", which was created by selecting Potato and changing Resistance Level from high to low and then selecting Economics and decreasing Costs from $1200 to $1100 per acre. You are spraying a systemic every two weeks in late June until harvest. (Note: When it rains, these sprays are delayed.)

- How does spraying compare to the use of high resistance potatoes in terms of profit?

- Are there other costs to consider?

Creating your own Late Blight management modules

You can construct multiple models in order to explore management strategies and the resulting losses or profits. (See the Start file to become familiar with this program.) Creating your own model involves simple steps such as:

- Go to File in the menu and choose New;

- Choose a weather file;

- Choose biological variables such as a variety of potatoes or levels of inoculum present and spore production;

- Choose a management strategies by selecting both frequency and types of applied sprays.

To advance through the growing season, click on the tab at the beginning of the bar at the bottom of the screen and drag it along to the desired date. You can stop multiple times. Use View to see the impact of different variables over the growing season including temperature, blight growth, and even residues from your spraying regime.

Using moderately resistant potatoes in a cool, wet season, develop your own management strategy for producing a yield that exceeds 200cwt/acre.

- Describe how you set up your model.

- Were you successful?

- What concerns do you have other than yield?

- Defend your management strategy in terms of preventing the development of resistant strains of late blight.

Optional: Designing a Novel Control Strategy

Crop scientists are always looking for new ways to control disease much like scientists working in the pharmaceutical industry. Designer drugs interrupt infection cycles by preventing initial contact between host cells and pathogens or by interfering with the production of new infective particles or cells. Similar molecular strategies are used in agriculture.

1. Choose a chemical used in crop management and explain its mode of action.

2. Write a brief proposal for a novel control method for Late Blight. First, explore the life cycle of the late blight microorganism, *Phytophthora infestans*, in potatoes. Then, describe what you will use, what part of the life cycle is being targeted, and how this method will interrupt the cycle or reduce spore production.

3. Crop scientists frequently use models and simulations to try to convince themselves and others of the reasonableness of their ideas. To demonstrate a new control method, they can build a model incorporating variables that show how the method interacts with the life cycle of the organism. Then they run simulations to produce results that will hopefully support their claims. You can use the *Late Blight Life Cycle* simulation program (Macintosh/Windows) located in the Potato Blight folder on the *Microbes Count!* web site to model your own control method. Brief instructions for using *Late Blight Life Cycle* are in the file *Modeling an Epidemic of Potato Fungus*.

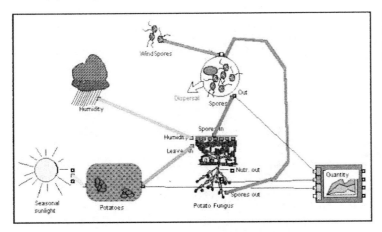

Figure 6. Screen shot of the default model in the *Late Blight Life Cycle* simulation, a resource available in the Potato Blight folder on the site.

You may want to familiarize yourself with the existing model first by exploring the scenario below:

• Scenario 1: Environmental conditions affecting the potato field can significantly alter the progress of the infection. (Infection occurs when wind-disseminated spores fall onto the leaf surface, germinate under cool, moist conditions and penetrate the cuticle.) To see what happens if the amount of rain is doubled, double-click the rain icon and enter in the dialog box that appears. Run the model to see the results.

Now consider what revisions need to be made to the model in order to show the impact of your method of control.

- List the model revisions.

- Make the changes and Run the model. Do the results of running this revised model support your approach to control? Explain.

Software Used in this Activity

Lateblight
> Phil Arneson and Barr E. Ticknor (Cornell University)
> > Platform Compatibility: Windows only

Late Blight Life Cycle
> Howard T. Odum (University of Florida) and Elisabeth C. Odum (Santa Fe Community College)
> > Platform Compatibility: Macintosh and Windows

Additional Resources

Available on the *Microbes Count!* web site at http://bioquest.org/microbescount

Software
> *Lateblight*

> *LateBlight Life Cycle*

Text
> A PDF copy of this activity, formatted for printing

> "Getting Started with *Lateblight*"

> "Getting Started with *Lateblight Life Cycle*"

Related *Microbes Count!* Activities

> Chapter 5: The Farmer and the Gene: A Case Approach to Bt Corn

> Chapter 9: Citrus Canker: Alternatives for Control

Unseen Life on Earth Telecourse

> Coordinates with Video X: Microbial Interactions

Relevant Textbook Keywords

> Hybrid, *Phytophthora infestans*, Resistance, Spores, Virus

Related Web Sites (accessed on 3/30/03)

> About Late Blight and *Phytophthora infestans*
> http://www.barc.usda.gov/psi/vl/lateblight.htm

Cornell-Eastern Europe-Mexico International Collaborative Project in Potato
Late Blight Control (CEEM)
http://www.cals.cornell.edu/dept/plantbreed/CEEM/

Microbes Count! Website
http://bioquest.org/microbescount

New Potatoes with Resistance to Late Blight
http://www.ars.usda.gov/is/pr/1998/981217.htm

Unseen Life on Earth: A Telecourse
http://www.microbeworld.org/htm/mam/is_telecourse.htm

References

Arneson, P. and B. E. Ticknor (2001). Lateblight. In *The BioQUEST Library Volume VI*, Jungck, J. R. and V. G. Vaughan, Editors. Academic Press: San Diego, California.Additional Resources

Fry, W. E. and S. B.Goodwin (1997). Resurgence of the Irish potato famine fungus. *Bioscience* 47(6):1-18.

Figure and Table References

Figure 1.   Courtesy Phil Arneson

Figure 2.   Courtesy Phil Arneson

Figure 3.   Courtesy Phil Arneson

Figure 4.   Screen capture of *LateBlight*

Figure 5.   Screen capture of *LateBlight* icon

Figure 6.   Screen capture of *LateBlight Life Cycle*

Making Sense of the Complex Life Cycles of Toxic Pfiesteria

Ethel D. Stanley, Elisabeth C. Odum, and Howard T. Odum

Video X: Microbial Interactions

Pfiesteria piscicida is found in the coastal and estuarine waters of the mid-Atlantic. This dinoflagellate was sensationalized in the news and by the publication of the popular book, *And the Waters Turned to Blood* (Barker 1997). Notorious for the production of a toxin linked to intermittent fish kills and human illness, *Pfiesteria* has been the focus of both public concern and scientific research in the past decade.

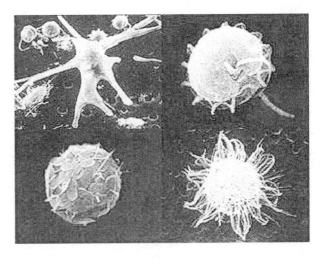

Figure 1. Amoeboid and zoospore forms associated with the life cycles of the dinoflagellate, *Pfiesteria piscicida.*

Scientists suspect that the normally benign microbe enters into a toxic phase when certain environmental conditions exist. Runoff from the land to the water is currently considered to be a major contributor to the development of toxic *Pfiesteria*. There is rapid growth of fish populations following increases in nutrients in the estuarine waters. The presence of abundant fish triggers a change in the normally benign dinoflagellate which begins to produce flagellated zoospores. This leads to the formation of toxic *Pfiesteria* that feed on the fish. Fish kills occur when the deadly toxin is released. Humans who are exposed to the toxin, such as local fishermen or scientists studying the phenomenon, can develop symptoms as well.

While *Pfiesteria* outbreaks are still actively investigated and data continues to be gathered, we can explore aspects of this complex life cycle through the following model and simulation.

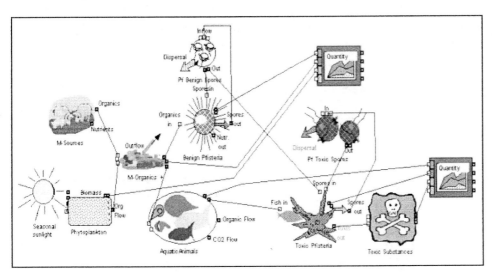

Figure 2. *Pfiesteria Life Cycle* model, created using the Extend modeling system.

Run the default model to simulate the population growth and decline of the non-toxic *Pfiesteria*. Levels of nutrient overflow fluctuate over time. Note the impact of excess nutrients on the microbial population in the graph below.

Figure 3. In the default model, the upper curve represents organics from runoff and the lower curve shows phytoplankton population levels. The non-toxic *Pfiesteria* population is shown between them. *Pfiesteria* increases noticeably following periodic increases in organics.

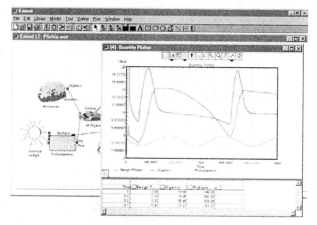

Figure 4. Note the rapid vertical spike of toxic *Pfiesteria* that occurs with the increase in aquatic life–specifically fish that are targeted by this microbe. This population decreases sharply indicating a "fish kill" event.

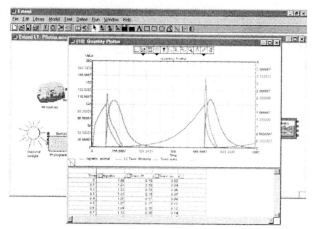

Explore the model by selecting a variable you think will impact the life cycle. You could choose to decrease the amount of organics in the runoff, increase the amount of inorganics in the runoff, increase the starting population of aquatic life, etc.

- Which variable will you change?

- Write down how you think the *Pfiesteria* populations will be affected.

Double click the icon for the variable you wish to change and enter a new starting value. Run the model after you make a change and note any changes you see in Pfiesteria populations.

- Were the results consistent with your predicted impact? If not, explain why.

  (Hint: If you wish to change a second variable, you may want to restore the first variable to its default or starting value.)

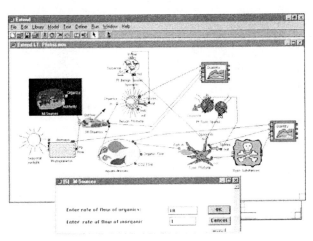

Figure 5. Increase the flow of organics/inorganics by doubling clicking the icon and then entering a new flow rate of 15. Click OK and then run the simulation to see the new data.

- Toxic outbreaks seem to occur in the aftermath of hurricanes that strike the mid-Atlantic states. Using the model, demonstrate possible links between these weather events and the resulting fish kills.

- Make an argument for supporting legislation in the Chesapeake Bay area which would restrict sites of pig farms. Generate data using the model to support your claims.

Optional activity: Life Cycle Controversy

Find a recent article on *Pfiesteria* in a scientific journal that includes a description of the life cycle. Provide the reference information for the article.

- Make a copy of this *Pfiesteria* life cycle or illustrate it yourself as it is explained in the article.

- List any problems that the authors faced trying to identify what was going on in the *Pfiesteria* life cycle.

Identify an online web site showing life cycle information available on the site. (Hint: *Pfiesteria* monitoring sites usually have this information.)

- Print out this life cycle of *Pfiesteria*:

- Compare the two life cycles. Are they the same? If not, how do they differ?

- How can you explain inconsistencies between the journal and online models? (Hint: Look for dates of publication, intended audience, etc. Do scientists always agree?)

- How do the life cycles compare to the *Pfiesteria Life Cycle* model you have been using?

- Models are necessarily simpler than the process being modeled. List at least three limitations in the *Pfiesteria Life Cycle* model.

Additional Resources

Available on the *Microbes Count!* web site at http://bioquest.org/microbescount

Software

Pfiesteria Life Cycle Model
Howard T. Odum (University of Florida) and Elisabeth C. Odum (Santa Fe Community College)

Platform Compatibility: Macintosh and Windows

Text

A PDF copy of this activity, formatted for printing

Related *Microbes Count!* Activities

Chapter 3:   Biosphere 2: Unexpected Interactions

Chapter 3:   Modeling Wine Fermentation

Chapter 7:   Microbiology of Stratified Waters

Chapter 10: Controlling Late Blight: Past, Present, and Future

Unseen Life on Earth Telecourse

Coordinates with Video X: Microbial Interactions

Relevant Textbook Keywords

Amoeboids, Dinoflagellate, Life cycle, Zoospores,

Related Web Sites

Chesapeake Bay *Pfiesteria* Monitoring
http://www.dnr.state.md.us/bay/cblife/algae/dino/pfiesteria/monitoring.html

Life Cycles of *Pfiesteria piscicida*
http://www.pfiesteria.org/pfiesteria/lifecycle.html

Microbes Count! Website
http://bioquest.org/microbescount

New NOAA Research Sheds Light on *Pfiesteria* Life Cycle
http://www.publicaffairs.noaa.gov/releases2002/june02/noaa02077.html

Pfiesteria piscicida, een probleem in Nederland: Mythe of realiteit?
http://www.rivm.nl/infectieziektenbulletin/bul97/rpfiest.html
http://www.rivm.nl/en/

Pfiesteria: Politics and Science Clash. Sean Henahan. Access Excellence
http://www.accessexcellence.org/WN/SUA12/pfiesteria298.html

Unseen Life on Earth: A Telecourse
http://www.microbeworld.org/htm/mam/is_telecourse.htm

References

Barker, R. (1997) *And the Waters Turned to Blood.* Simon & Schuster. NY

Burkholder, J. M. and H. B. Glasgow (2001). History of Toxic Pfiesteria in North Carolina Estuaries from 1991 to the Present. *Bioscience. 51 (10):* 827-841.

Odum, H. T. and E. C. Odum (2002). *Pfiesteria.* Contact the BioQUEST Curriculum Consortium for information.

Figure and Table References

Figure 1.   Courtesy of The National Institute of Public Health and the Environment (RIVM) (http://www.rivm.nl/en/), The Netherlands http://www.rivm.nl/infectieziektenbulletin/bul97/rpfiest.html

Figure 2.   Screenshot from the *Pfiesteria Life Cycle* model

Figure 3.   Screenshot from the *Pfiesteria Life Cycle* model

Figure 4.   Screenshot from the *Pfiesteria Life Cycle* model

Figure 5.   Screenshot from the *Pfiesteria Life Cycle* model

Measles in Nakivale Refugee Camp: Epidemiology in Action

Emily Smith, K.C. Keating and Marion Field Fass

Video X: Microbial Interactions

Measles is a childhood disease that is no longer a major threat to most American children. From a global perspective, however, measles is still a devastating killer. According to the World Health Organization measles causes approximately 900,000 deaths each year, 98% of which occur in developing countries (World Health Organization website). In this activity we will develop models of the effectiveness of strategies to halt the spread of measles in a refugee camp in the African country of Uganda and explore the potential impact of other infectious disease control strategies in this population.

In order to develop an accurate model of the impact of a program to control an outbreak of measles, we must understand the natural history of the disease. Measles is a viral disease that is easily spread through casual contact with an infected person. In 1963 a vaccine for measles was licensed in the United States. At present, most American children receive the vaccine as part of the MMR (measles, mumps, and rubella) series, a standard childhood immunization, and the incidence of the disease in the US has fallen dramatically. Before the widespread use of this vaccine thousands of children each year suffered seizures, brain damage, deafness, or death due to measles infections.

Figure 1. African children.

Background

The symptoms of measles are influenza-like, including dry cough, fever, and malaise. In addition, there is a characteristic maculopapular rash and Koplik's spots in the mouth. Measles is caused by a virus of the family *Paramyxovirinae*, genus *Morbillivirus*. The single stranded negative-sense RNA virus is carried only by vertebrates: humans, wild monkeys, and some rodents. There is no vector involved in disease transmission. However, since infected individuals are infectious before they show definitive symptoms, a single case can easily lead to an outbreak.

Measles spreads rapidly through a population because the virus reproduces in the cells of the nose and throat lining and can be transmitted by coughing, sneezing, or even talking to a susceptible person. After infection occurs, the virus has an initial growth phase of 4-6 days. It then travels to the liver and spleen where further replication occurs. After 10-12 days the virus invades the cells of the epithelium of the eyes, lungs, and stomach and the individual begins to show symptoms. The characteristic rash appears by the 14-16th day, a few days after onset of flu-like symptoms, and is a result of the reaction to the virus by the epithelial cells (Centers for Disease Control website). An infected individual is contagious from roughly four days before the skin rash appears to four days after, and can easily spread the virus to 90% of their close contacts (Centers for Disease Control website). There is no treatment for measles itself, only for discomfort and associated conditions such as encephalitis, ear infections, or pneumonia. Most individuals recover but complications such as brain damage, deafness, blindness, and death can occur. Measles is more severe in malnourished children, particular those with Vitamin A deficiency, and the fatality rate can be as high as 25% (Centers for Disease Control website). When an individual recovers, antibodies should protect against re-infection for life.

The best way to prevent the spread of measles in a community is to establish a high level of immunization. Ideally, 90% of the population should be immunized to keep the number of susceptibles very low. The most commonly used vaccine is a live attenuated virus. It is highly effective in providing immunity–approximately 95% of children achieve immunity after receiving the first dose. A second dose is administered to cover those few who do not develop initial immunity and as a booster shot for everyone else (Centers for Disease Control website). Globally, countries with high rates of immunization experience very few cases of measles.

The Problem

In the spring of 2000, the Nakivale Refugee Camp in Southwestern Uganda experienced an outbreak of measles (Keating, 2000). Among the factors involved in this outbreak were low immunization levels and poor health of arriving refugees. The Nakivale Camp housed refugees from eight different African nations: The Democratic Republic of Congo (DRC), Kenya, Rwanda, Burundi, Somalia, Ethiopia, Eritrea, and Sudan. These groups were divided, or divided themselves, into five different settlement zones based on nationality and arrival date. In all, more than 5,000 people lived in the Nakivale Camp.

The challenge faced by local health workers who took on the task of fighting the outbreak was complicated. What would be the best course of action in the face of an outbreak of highly contagious disease?

Using the software program *Epidemiology*, shown in Figure 2 below, you can investigate some of the strategies that could be used to limit the impact of a measles outbreak among people living in the Nakivale Refugee Camp. Information on using *Epidemiology* is available on the *Microbes Count!* site. You might also want to look at the "Epidemiology: Understanding Disease Spread" activity in Chapter 11 for a more detailed example of using the *Epidemiology* application.

Figure 2. A screen shot of the control window in *Epidemiology*.

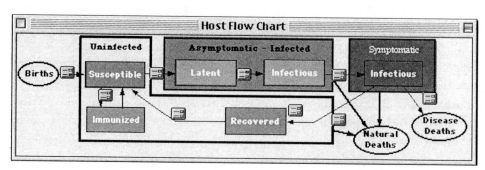

Part 1: The Basic Model

When you model the measles outbreak in Nakivale what characteristics of the population, the disease, and the environment will you need to consider? For this activity you will be using an *Epidemiology* model file called "MeaslesNakivale". "MeaslesNakivale" is one model of the spread of measles through the Nakivale population. You may want to explore the settings used in the "MeaslesNakivale" model to be sure that you understand how the real-life situation in the Nakivale camp has been translated into the computer model.

Begin your investigation by starting a new problem in *Epidemiology* using the "MeaslesNakivale" file (located in the *Epidemiology* folder on the *Microbes Count!* web site.) The entire course of a measles infection takes only a couple weeks, so you should assume the time units represent days and make other setting decisions based on these units.

The settings for "MeaslesNakivale" are summarized in Table 1. These settings are based on a population of 5000 individuals, of whom 50% are under 18. We will assume that those over 18 are immune to measles through prior infection or immunization.

Table 1: Parameter settings for the MeaslesNakivale model.

Parameter	Setting
Total population	5003
Susceptible	2000
Recovered	2500
Infectious	3
Immunized	500
Probability of transmission	0.5 per contact
Contact rate	Density dependent 5/1000/time period
New cases	1 per 50 time periods
Average duration of latency	10 time periods
Asymptomatic infection	3 time periods
Duration of infection	6 time periods
Disease mortality	0.02 per time period
Natural mortality	0.002 per time period
Birth rate	0.002
Duration of immunity	1000
% of newborns immunized at birth	0%

- Using the "MeaslesNakivale" scenario, investigate the movement of measles through this population for 100 days. On the Control Menu you can choose "Go Until" to see the spread of the epidemic in steps of 5 or 10 days.

Based on the "MeaslesNakivale" model and what you know of the measles spread, map out your plan for vaccinating the residents of the Nakivale camp. You can schedule regular weekly clinics to immunize babies and young children at risk, or hold a mass immunization day to reach as many people as possible.

To model a mass immunization day, go to the Settings menu and pull down Current State. You can subtract people from Susceptible category and add them to Immunized. When you return to the graphs they will adjust to account for your actions.

To model the immunization of babies, click on the box between susceptible and immunized on the Host Flow Chart, and set the effectiveness level of your campaign.

- How important is it to act quickly in response to this outbreak?

- What are the effects of different control strategies at 30, 50 and 100 days?

Part 2: Modifying the model

One of the most difficult aspects of the Nakivale camp outbreak to model is the behavior of the refugees themselves. In the "MeaslesNakivale" model, measles transmission is Density Dependent because spreads to more people in more crowded conditions. But how do we determine how many people a refugee in the Nakivale camp comes into contact with in a given day? Of course, this is nearly impossible to know for sure, and is likely to be different for each individual. The refugees separate themselves based on nationality of origin and presumably by other less obvious features as well, but this does not necessarily mean they have no contact at all with other groups. The estimate used in the "MeaslesNakivale" model is based on the number of latrines per person in each group. The Latrine Survey (Table 2) was originally done as part of a project to estimate the priority of preventative health measures in each Zone, but the results can also be used to estimate the number of contacts each person has per day. Table 2 shows the data collected in the Latrine Survey.

- Perform an analysis of the pit latrine variable by changing the number of contacts an individual has each day. (To modify the number of contacts, click on the Transmission Parameters box (see Figure 2) or choose Model:Transmission Rate under the Settings menu.) Remember that the number is contacts per 1000 people in the population.

If the number of people using each latrine actually reflects the number of contacts of each infected person, do the Zones with high people to latrine ratios have a significant disadvantage in a measles outbreak?

Zone	Nationality Represented	Latrines	People	Ratio People:Latrine
Old Rwandese	Rwandese Kenyan	41	1202	29.32:1
New Rwandese	Rwandese	27	1358	50.30:1
Congo	Congolese	40	612	15.30:1
Zaire	Congolese	59	734	12.44:1
Somali	Somalis Ethiopian Eritrean Burundian Sudanese	56	1006	17.96:1
New Arrivals	Rwandese	7	370	52.86:1

Table 2. Nakivale Refugee Camp Latrine Survey Results Spring 2000

- If the number of contacts per person per day can be decreased by increasing the number of usable latrines, would a 'latrine-building' campaign be more effective in curbing a measles outbreak than an immunization program? Over the course of 100 days, compare and contrast the results of each campaign. Test the effectiveness of your strategies by introducing new infected individuals? How do these strategies compare?

- Do you feel comfortable with this method of determining number of contacts? How else might you estimate such a number? Do cultural or personal behavior patterns influence any of the other variables in the *Epidemiology* model? In the outbreak at Nakivale, the highest rate of measles cases actually occurred in the Congolese population where the Latrine ratio was lowest. Keating (2000) suggests that people changed their behavior when it was known that cases of measles had been reported. People similarly change behavior as a result of health education when outbreaks occur. Provide other examples and explain how they might affect disease spread in this population.

Part 3. Expanding the model

You can use this model to investigate the impact of other infectious diseases in a refugee camp, where population density is high and the health of the inhabitants has been compromised by hunger, stress and other diseases. Explore the impact of cholera and other diarrheal diseases, or tuberculosis, or upper respiratory infection in this setting.

For further investigation of these questions, or to expand on this model, select the Settings menu, and choose the option Define Model. To accurately reflect the refugee camp at the beginning of an outbreak, go back to the Settings menu and choose Current State. Here you can set the population lower and reflect the number of cases initially present, as a very small number of cases (3-5) should be enough to represent the beginnings of an outbreak. From there, you can change the other settings to model different vaccination campaigns, entrance of a large number of new susceptibles, or changes in other variables.

- The United States has implemented a vaccination program successful enough to claim that all endemic measles has been eliminated. Is this a possibility for the people of Uganda? If so, what would be a plan of action for moving towards this goal? If not, what are the factors that make this goal difficult to achieve?

Software Used in this Activity

Epidemiology

Dan Udovic and Will Goodwin (University of Oregon)

Platform Compatibility: Macintosh

Additional Resources

Available on the *Microbes Count!* web site at http://bioquest.org/microbescount

Software

Epidemiology

The "MeaslesNakivale" model file for *Epidemiology*

Text

A PDF copy of this activity, formatted for printing

Keating, K. C. (2000). *Measles in the Nakivale Refugee Camp: Epidemiology and Control of April 2000 Outbreak.*

Related *Microbes Count!* Activities

Chapter 11: Epidemiology: Understanding Disease Spread

Chapter 12: Vaccine: Experimenting with Strategies to Control Infectious Disease

Unseen Life on Earth Telecourse

Coordinates with Video X: Microbial Interactions

Relevant Textbook Keywords

Asymptomatic infectious phase, Growth phase, Immunity, Latent phase, Vector

Related Web Sites (accessed on 2/25/03)

Background Information about Measles
http://www.thedoctorsdoctor.com/diseases/measles.htm

Microbes Count! Website
http://bioquest.org/microbescount

Unseen Life on Earth: A Telecourse
http://www.microbeworld.org/htm/mam/is_telecourse.htm

References

Centers for Disease Control and Prevention
http://www.cdc.gov/nip/publications/pink/meas.pdf

Keating, K. C. (2000). *Measles in the Nakivale Refugee Camp: Epidemiology and Control of April 2000 Outbreak.* Original Research Project through the School for International Training, Kampala, Uganda. For reprint of the article, contact BioQUEST.

The Doctor's Doctor
http://www.thedoctorsdoctor.com/diseases/MEASLES.HTM

Udovic, D. and W. Goodwin (2001). Epidemiology, In *The BioQUEST Library Volume VI.* Jungck, J. R. and V. G. Vaughan, editors. Academic Press: San Diego, CA.

World Health Organization
http://www.who.int/measles/app/docs/Measles_brochure.pdf

Bibliography

Medecins Sans Frontiers (1997). *Refugee Health: An Approach to Emergency Situations.* MacMillan: London.

Figure and Table References

Figure 1.   Marion Fass

Figure 2.   Screen shot from *Epidemiology*

Table 2.   Keating, K. C. (2000)

Chapter 11

Activities for Video XI: Human Defenses

Physicians recognize key symptoms presented during patient examinations and further the reliability of their diagnoses by requesting confirmatory clinical tests for immunological responses that are not otherwise easily detected. In some cases, human defenses to microbial infection are probed by clinical testing procedures that identify specific patient responses to infection and disease.

In this unit we can:

- offer a diagnosis using patient data and results from selected clinical testing and ascertain if the process we used is efficient and effective,
- identify lymphocytes in a clinical laboratory simulation of blood cell counts, and
- build epidemiological models of different diseases, design strategies for disease control, and test the effectiveness of these strategies on virtual populations.

Identifying Immune Responses

AuntieBody is an online IMMEX problem-solving simulation for testing your clinical skills and knowledge of the immune system. The interactive simulation offers patient information and asks for a sensible and efficient diagnosis. Clinical tests may be selected and the results assessed before submitting a diagnosis.

Why Count Types of White Blood Cells?

How can we make use of complex cellular level responses in the human body to microbial infections and other disorders? Why is it important to differentiate between white blood cells in a blood sample and keep a record of their numbers? Improve skills at cell identification and explore these questions with the program Cell Differentials.

Epidemiology: Understanding the Spread of Disease

Factors that influence disease spread throughout populations can be explored with the program Epidemiology. Both population and disease characteristics can be modeled over different time periods. The Susceptible-Infected- Recovered (SIR) model enables us to make predictions based on significant variables such as the flow of new susceptibles in to the population, transmission rates, disease deaths, and the duration of the disease. Ebola is used as a model organism and epidemiology is presented from both a microbiological and social perspective.

Identifying Immune Responses

Ethel D. Stanley

Video XI: Human Defenses

Pathogens, allergens, or tumor growth can trigger the production of specific antigens. Immune cells and immunoglobulins react to the presence of these antigens in the body. Physicians use the specificity of immune responses to help determine the causative agent for problems their patients are reporting.

You have an opportunity to use your knowledge of immune responses and to practice your diagnostic skills in an online simulation. The IMMEX AuntieBody software provides:

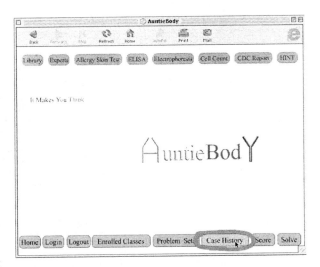

Figure 1. The AuntieBody web site.

- A patient case history;

- Access to diagnostic techniques such as blood tests, electrophoresis, ELISA (enzyme-linked immunosorbant assay), and allergy skin tests;

- Background information on the technical and medical terminology;

- Advice from professionals who use the techniques;

- Results for your patient; and

- An automatic report for tracking the medical expenses you generate.

Getting Started with Online IMMEX on the Web

Go to http://bioquest.org/microbescount on your internet browser.

Choose *Identifying Immune Responses* from the list of activities. Follow the directions to access the IMMEX AuntieBody problem solving simulation.

In the AuntieBody window, click on Case History to get a randomly selected patient to diagnose.

Read the Case History. Near the top of the page is a row of button for the different resources that are available to help you explore the problem further. Depending on your case, choose resources that you think will help you make an accurate diagnosis.

Resource Example: Cell Count

If you choose Cell Count, you will see three different Cell Count tests that you can order. Click on Differential Count. You are advised that the Differential Count will cost $30, leaving you with $970 for further tests or consultations. Remember that you are trying to diagnose the ailment efficiently (fewest tests to keep patient costs down), but with accuracy. You will have to decide whether you want to order this test. You can use the Library and Expert options to further your own understanding of the test. If you click Yes the test results will appear.

Figure 2. Test results for the differential count.

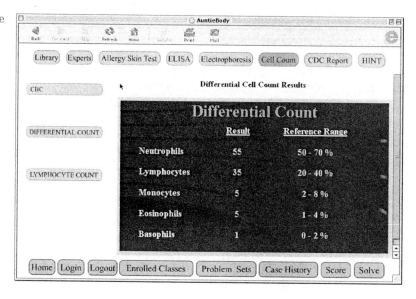

Continue with this case until you feel ready to make a diagnosis. You may want to familiarize yourself with the latest CDC report on disease statistics or to check to see if your patient's symptoms are similar to those reported in a recent outbreak.

You will be given an option to finalize a diagnosis and then can check to see if you are correct.

Reflect on your experience by addressing the questions below:

- What might account for the symptoms your patient is exhibiting? Is a microbial cause likely?

- What tests did you order? Explain what you were looking for with each test.

- Did any of these tests allow you to rule out potential causes? How?

- Did you use resources other than tests? List each and explain which were the most helpful.

- Were you successful in making a diagnosis?

- If confronted by a similar patient what would you do differently? Why?

Software Used in this Activity

AuntieBody
IMMEX Project
Ron Stevens (UCLA), Mel Stave (Ulysses S. Grant High School), and Adrian Casillas, MD (UCLA)
Platform Capability: Macintosh and Windows

Web Resources in this Activity

IMMEX Project at UCLA
http://www.immex.ucla.edu

Additional Resources

Available on the *Microbes Count!* web site at http://bioquest.org/microbescount

Text

A PDF copy of this activity formatted for printing

Getting Started with IMMEX on the Web

Related *Microbes Count!* Activities

Chapter 4: Exploring HIV Evolution: An Opportunity for Research

Chapter 11: Why Count Types of White Blood Cells?

Chapter 12: Souvenirs: Investigating a Disease Outbreak

Unseen Life on Earth Telecourse

Coordinates with Video XI: Human Defenses

Relevant Textbook Keywords

Allergen, Antigen, Basophil, Electrophoresis, Eosinophil, Immunoglobulin, Lymphocyte, Monocyte, Neutrophil, Pathogen, Western blot

Related Web Sites (accessed 4/25/03)

Centers for Disease Control and Prevention
http://www.cdc.gov/

Microbes Count! Website
http://bioquest.org/microbescount

NIH: Health Information
http://health.nih.gov/

United States Department of Health and Human Services: Reference Collection
http://www.os.dhhs.gov/reference/index.shtml

Unseen Life on Earth: A Telecourse
http://www.microbeworld.org/htm/mam/is_telecourse.htm

Figure References

Figure 1. Screen shot of AuntieBody

Figure 2. Screen shot of AuntieBody

Why Count Types of White Blood Cells?

Ethel D. Stanley and Donald Buckley

Video XI: Human Defenses

Phagocytic white blood cells are essential to non-specific defense against invading microbes. Other kinds of white blood cells provide specific defense against a microbe that has been encountered before. Making counts of the kinds of white blood cells present can help us decide which microbial invaders might be at work. (See Table 1: Summary Information for White Blood Cells at the end of this activity.)

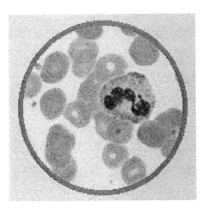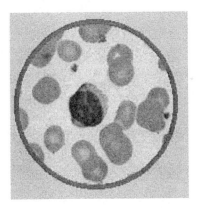

Figure 1. The larger, stained cells observed in these blood smears are a neutrophil (left) and lymphocyte (right)

Differential white blood cell counts are routinely ordered for patients and do provide diagnostically valuable results. These "cell differentials" are usually done on automated flow cytometry instruments using a sample of blood. Approximately 10,000 white blood cells (WBCs) are classified on the basis of size and peroxidase staining. Neutrophils, monocytes and eosinophils are peroxidase positive. Lymphocytes are peroxidase negative. Basophils are identified using two angle light scattering on their resistance to lysis.

However, manual cell differentials are also necessary. In about 20% of the samples, technicians visually inspect stained blood smears not just for WBC counts, but for specific morphological abnormalities. As you can see in Figure 1 above, white blood cells have similar features. WBC's also vary due to developmental changes. Making accurate cell counts from a stained blood smear can be difficult at first, but it produces valuable data for health professionals.

Using Cell Differentials

Cell Differentials was developed to provide real-time practice in distinguishing between types of white blood cell. Using this program before you go to the lab to do cell counts should be helpful. You can monitor your skill level over time. *Cell Differentials* provides opportunities for extended visual practice with instant feedback. You are encouraged to repeat the experience until you have become proficient at identifying white blood cell types.

Getting started

- Click the switch in the top-right of the screen. A blood smear featuring an unknown white blood cell among red blood cells will appear in the microscope field (Figure 2).

Figure 2. A screen shot of Cell Differentials showing the selection of the Neutrophil key (cursor on lower left key) in response to the identification of the white blood cell shown in the field on the right.

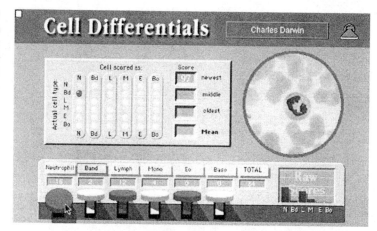

- Identify the cell type and then click on the appropriate key on the clicker at the lower left. The number of each cell type is recorded in a small field above each key and displayed graphically in the histogram on the right side of the clicker.

- If the cell is correctly identified, a new image will be provided. If an incorrect cell identification is made, the correct answer is highlighted. Then click anywhere to continue the exercise.

- A summary table is shown on the top-left of the screen. Clicking on the red or green circles brings up a dialog box that provides more details about your recognition of those particular cell types.

- While making manual cell differential counts, a lab technician notices that he is hitting more keys on the left of counter with the latest sample. What explanation can you offer for this "left shift"?

Laboratory Forensic Investigation

As an extended lab exercise, do differential white blood cell counts on blood sample unknowns provided by your instructor.

1. Record the WBC count for each slide.

2. Is it in the normal range?

3. Suggest potential explanations for these results.

Building Your Own Visual Data Set

Option 1: Obtain 10 images of white blood cells from the internet or by scanning print materials. Inspect each of these images and decide whether or not this is or is not a good representative image. Explain.

Option 2: Locate at least 5 images of abnormal white blood cells. Inspect the image and describe features that can be used to identify the cells as abnormal.

Table 1: Summary Information for White Blood Cells

Names of White Blood Cells	Neutrophils polymorpho-nuclear or segmented neutrophils PMNs, segs, or polys	Band Neutrophil bands	Lymphocytes lymphs	Monocytes monos	Eosinophils eos	Basophils basos
Visual ID	multi-lobed nucleus in light stained granular cell	large band nucleus in light stained granular cell	large nucleus nearly fills clear cell of smaller size	U or kidney bean shaped nucleus in clear larger cells	Two-lobed nucleus in cell with large pink granules	numerous granules blue to purple, often hiding nucleus
% WBC in adults	50-70%	1-4%	15-35%	2-10%	1-4%	0.5-1%
Decrease in blood cell differential count may indicate:	typhoid, advanced sepsis, viral infections, such as influenza, red measles or hepatitis	bone marrow dysfunction	severe stress, cancer treatments decreases in T4 cells to measure AIDS	depleted in overwhelming bacterial infection	acute and chronic inflammation, stress, drugs: steroids	hemolytic anemia or chicken pox
Increase blood cell differential count may indicate:	bacterial infection, leukemia, tissue death, as in burns or gangrene	shift to the left (hit more left keys on the counter) usually indicates infection	viral infection, such as infectious mononucleosis, pertussis	chronic infections, tuberculosis, cancer, and disorders of the immune system	drug effect, allergy or parasitic infestation of the gut, malaria	hyper-sensitivity reactions, drugs, myeloproliferative disorders
Functions:	phagocytic granulocytes produce toxic substances that kill and digest invaders, also attract eosinophils	immature neutrophil	develop into B and T cell types, plasma cells produce antibodies, memory cells direct immune response; . attract eosinophils	large phagocytic white blood cell which, when it enters tissue, matures into macrophages (phagocytes)	granulocytes: release chemicals as well as enzymes that decrease inflammatory reactions	granulocytes: release chemicals (e.g. histamine; heparin) develop into mast cells, allergy and inflammatory reactions

Additional Resources

Available on the *Microbes Count!* web site at http://bioquest.org/microbescount

Software

Cell Differentials: A White Blood Cell Identification Exercise
Don Buckley, Deborah Clark (Quinnipiac University), Karen Barrett, Lynn Gugliotti, and JoAnne Morrica (University of Hartford)

Text

A PDF copy of this activity, formatted for printing

"Table 1-WBC Key Charact.pdf" – the Summary Information for White Blood Cells table in a larger format for printing

Cell Differentials: An Introduction

Relevant Textbook Keywords

Morphology, White Blood Cells (WBC)

Related *Microbes Count!* Activities

Chapter 11: Identifying Immune Responses

Unseen Life on Earth Telecourse

Coordinates with Video XI: Human Defenses

Related Web Sites

Differential White Cell Count - Theory
http://www.medicine.mcgill.ca/physio/vlabonline/bloodlab/whitediff.htm

Microbes Count! Website
http://bioquest.org/microbescount

Unseen Life on Earth: A Telecourse
http://www.microbeworld.org/htm/mam/is_telecourse.htm

White Blood Cell Count and Differential
http://www.ahealthyme.com/topic/topic100587682

White Blood Cell Differential Count
http://www.labtestsonline.org/understanding/analytes/differential/sample.html

References

Buckley, D., D. Clark, K. Barrett, L. Gugliotti, and J. Morrica. (2001). Cell Differentials: A White Blood Cell Identification Exercise. In *The BioQUEST Library Volume VI*, Jungck, J. R. and V. G. Vaughan, Editors. Academic Press: San Diego, California.

Figure and Table References

Figure 1.   Screen shot from *Cell Differentials*

Figure 2.   Screen shot from *Cell Differentials*

Epidemiology: Understanding Disease Spread

Marion Field Fass

Video XI: Human Defenses

Epidemiology is the study of the distribution of diseases in populations. Epidemiology embraces the molecular and genetic factors that influence susceptibility to disease, the biochemical factors that cause some agents to be more infectious than others, and the broad characteristics of the physical and social environments in which diseases spread. It enhances the microbiological perspective on pathogens by exploring the reasons why diseases reach "epidemic" proportions in some situations and just fade away in others. Epidemiology is a tool used by public health practitioners and clinicians alike to determine the strategies to control diseases. The tools of epidemiology have also been used by social scientists to study "outbreaks" of ideas and behaviors. (Gladwell, 2002)

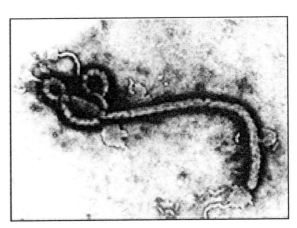

Figure 1. Electron micrograph of Ebola Zaire virus. Taken in 1976 by Dr F.A. Murphy, now at UC Davis, then at the CDC, this is the first photo ever taken of the virus (http://www.abc.net.au/science/news/stories/s59403.htm)

Epidemiologists study qualities of the agent of disease, the host, and the environment to characterize outbreaks of disease and to predict their impact. Their models of disease make assumptions about how microbes and their hosts behave in given situations. In order to develop an accurate prediction of disease spread, an epidemiologist will develop multiple models based on a range of assumptions about the virulence of the agent, transmissibility, flow of new susceptible hosts into the population, and rate of recovery of infected individuals. By changing the assumptions of the model, epidemiologists can experiment in ways that aren't feasible in large, human populations.

In this activity you will use a computer program called *Epidemiology* to simulate the spread of a disease through a population. By manipulating values such as death rates, birth rates, the probability of disease transmission, and initial population characteristics, and then using the simulation to see how population characteristics change through time, you can test the impact of treatments and behavioral and environmental changes over time.

What Are the Factors that Affect Disease Spread?

Work with a group of three other students to develop a concept map to demonstrate the spread of a disease in a population. You may choose a real disease or design a disease of your own. The Centers for Disease Control and Prevention (CDC), headquartered in Atlanta, Georgia, track infectious and chronic disease in the United States and other countries. They publish brief reports on outbreaks and trends in the newsletter, *Morbidity and Mortality Weekly Reports (MMWR)*, and more detailed analyses of infectious diseases in the on-line journal, *Emerging Infectious Diseases*. Both of these are available at the CDC website (http://www.cdc.gov) and would be good sources of information about the characteristics of outbreaks of infectious diseases. Some other resources are listed in the Additional Resources section at the end of this activity.

> • What factors determine the spread of the disease in the community that you have chosen? How are people exposed to the pathogen?
>
> • How is the spread of "your disease" affected by the community in which the outbreak has occurred? Is everyone at equal risk of infection?
>
> • Is "your disease" one where someone who has been infected and recovers never gets the disease again, or can they be reinfected when exposed again, as with gonorrhea or the common cold? How long does immunity last?

One way of modeling diseases spread uses estimates of numbers of susceptible, infected and recovered individuals in a population. This SIR (Susceptible, Infected, Recovered) model also must take into account the behavior of those individuals in their environment, and the effectiveness of the pathogen. Did your model account for all of these factors?

Using *Epidemiology* to Model Disease Spread

Now that you have developed an understanding of the factors that need to be considered when predicting disease spread, you are ready to use the *Epidemiology* simulation with a simple disease model; later on you will use *Epidemiology* to create a more complex model that allows you to investigate the effects of vector-borne diseases, and to estimate the impact of immunizations

Overview

Our simple model asks you to input data on six characteristics of the disease you are studying and the population that it is affecting. These characteristics are:

- Transmission rate,
- Duration of infection,
- Duration of immunity,
- Birth rate,
- Death rate, and
- Disease death rate.

In *Epidemiology* you can change the values of the factors that influence these characteristcs using the Flow Chart window (Figure 2). Clicking on one of the small, grey boxes in the flow chart will open a dialog box in which you can set the values of the parameters for that part of the model. Alternatively, you can also display the dialog boxes by clicking on the Model item under the Settings menu. This will bring up a menu listing the dialog boxes for the various factors that influence the model.

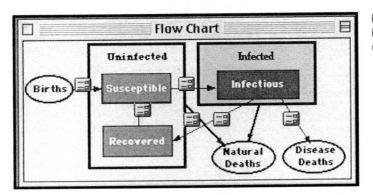

Figure 2. Screen shot of Epidemiolgy Flow Chart used for changing model characteristcs.

The "Epidemiology Users Manual" on the *Microbes Count!* site has more detailed instructions on using Epidemiology. The instructions and screen shots below and in the "Users Manual" refer to the Macintosh version of Epidemiology (included on the *Microbes Count!* site). A web-based version is also available (http:// darkwing.uoregon.edu/~bsl/epidemiology/index.html). The controls and graphs in the web-based version may look slightly different from the images pictured here but the basic functionality is similar.

In *Epidemiology*, the results of a simulation are displayed in two graphs. The "Death vs Time" graph (Figure 3a), shows the natural death rate and the disease death rate. The "Population" graph (Figure 3b) records changes in the number of susceptible, infectious, and recovered individuals over time. (Your graphs will not necessarily look exactly like these.) You can switch between a graphical and a spreadsheet display by clicking on the Table icon in the upper left. Data can also be displayed as either total numbers or as a percentage of the total (click on the "%" icon in the upper left.)

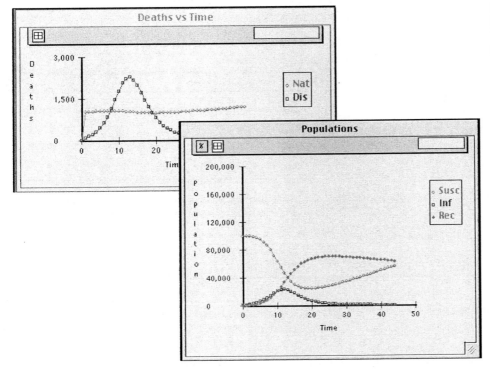

Figure 3. Results of a simulation in Epidemiology: (a) the Death versus Time graph and (b) the Population graph.

Modeling Ebola

The illness known as Ebola Hemorrhagic Fever was first described in a rural area in Zaire (now the Democratic Republic of Congo) in 1976. *The Hot Zone*, Richard Preston's story of Ebola in a lab in Reston, Virginia brought the virus to popular attention and established its horror. Ebola Hemorrhagic Fever is caused by the Ebola virus, a member of the family Filoviridae. The incubation period is 2-21 days and the mortality rate in large outbreaks ranges from 53% to 88%. (Special Pathogens Branch, CDC).

The epidemiologists' report of the first outbreak of Ebola is available on the website for the Prince Leopold Institute of Tropical Medicine in Belgium (Breman, 1978).

The Ebola model we will use is loosely based on this 1976 outbreak. During this outbreak, much of the spread of the disease occurred in people who were patients at the small local hospital, so in our model, the hospital is our population. The population size has been set to 300 to model a small community characterized by the patients and staff at the hospital and the people who visit the hospital and are in contact with patients and health care workers. At the beginning of the outbreak, when people do not yet understand how the disease is spread, both the number of contacts and the probability of transmission given contact are high. Mortality for Ebola is about 25% per week.

The settings for our model are summarized in Table 1. To examine these settings, open the *Epidemiology* model file called "Ebola Zaire". This file is located in the Epidemiolgy folder on the *Microbes Count!* site. Once the model has loaded, click on the Settings menu and choose Model:Transmission Rate. To see the Population settings, choose Current State... under the Settings menu. For the mortality rate, choose Disease Death Rate.

Table 1. Parameter settings for the "Ebola Zaire" model.

Parameter	Setting
Total Population	301
Susceptible	300
Infectious	1
Recovered	0
Probability of Transmission	0.5 per contact
Contact Rate	Density Dependent
Average Number of Contacts	15 per time interval, per 1000 pop.
Mortality Rate (per capita disease mortality rate)	0.25 per time interval

For this model we have defined the unit of time to be weeks. (Later on, when you build your own models of disease, you may choose to define the units of time to be days or weeks or even months, depending on the natural history of the disease.)

Now run this scenario for 30 weeks to see what happens to the population. Under the Control menu, choose Go Until... or Go For... and select 30.

At this point you may want to save your model so that you can go back to it later. Choose Save As... (*not* Save) under the File menu, give your model a name, and click the Save button. Be sure to take note of the location where the file is saved.

Now let's model the effect of intervention to stop the outbreak. First reset your model so that you can rerun the simulation. To reset the model, choose Reset from the Control menu or click on the button on the far right of the small control bar at the top left of your screen. First, run the simulation with the original settings for five weeks, the time required to get the epidemiologists on the scene to help stop the outbreak. (Under Control, choose Go Until... and select 5.) As they analyze the situation, the epidemiologists make recommendations to quarantine the hospital and to halt the reuse of syringes. In our model, this reduces the contact rate to five, and the probability of transmission to 0.1. Make these changes in the model by changing the appropriate parameters. (Pull down the menu for Settings and choose Model. Under Model, choose Transmission Rate.)

Run your new scenario for 30 time units. Save your results under a new name so that you don't overwrite your previous results.

Below is a set of data from the first recorded outbreak of Ebola in Zaire, now called Democratic Republic of Congo, in 1976 (Breman, 1977).

Table 2. Data from the first recorded outbreak of Ebola (Breman, 1977).

Number of Cases of Ebola by Date and Mode of Transmission			
Dates	Person to Person	Syringe	Both
September 1-3	0	3	2
September 4-6	3	3	1
September 7-9	4	7	0
September 10-12	5	15	1
September 13-15	14	13	0
September 16-18	7	10	3
September 19-21	9	14	2
September 22-24	26	18	2
September 25-27	19	3	3
September 28-30	11	1	0
October 1-3	12	1	0
October 4-6	12	1	2
October 7-9	7	0	1
October 10-12	10	1	0
October 13-15	3	1	0
October 16-18	0	0	0
October 19-21	3	0	0
October 22-24	1	0	0

- Does the "Ebola Zaire" model accurately describe the progress of this outbreak?

- How would you change the parameters to more accurately reflect what happened?

- Can you use your model to predict the impact of modern control measures, or quarantine on the spread of the disease?

Designing Models of Other Diseases

Now that you are familiar with the program *Epidemiology* and the factors that contribute to disease spread, you can develop models that enable you to experiment with the factors that influence disease spread and with the impact of changes in behavior and treatment.

The Centers for Disease Control (CDC), the World Health Organization (WHO) and the Program for Monitoring Emerging Diseases (ProMED) can provide information on the epidemiology of various diseases. Some other resources are included below and in the Additional Resources section at the end of this activity.

For this activity,

1) Develop a hypothesis about how changes in characteristics of interest would affect disease spread for one of the conditions listed below or another of your choice.

2) Test your hypothesis by generating a set of graphs and charts in *Epidemiology* to provide data.

3) Write a report or prepare a poster to describe the impact of these changes and how your predictions would help a policy maker or health professional to limit the spread of disease.

Below you will find some suggested epidemiology scenarios. You can start with the simple model of disease spread that is the default model in *Epidemiology* (choose New from the File menu) and modify it to fit the characteristics of your scenario. This is the way the "Ebola Zaire" model was created. You may also want to add factors such as vectors and immunity to create a more realistic model. To add the impact of vectors or to model the role of immunization, go to the Settings menu and select Define Model.

- Can you model the spread of colds through a population in winter? Remember that the protection from re-infection is negligible, either because there is no acquired immunity, or because there are so many "cold viruses" rhinoviruses out there. *Epidemiology* allows you to introduce new infected people into the population through the Model options under the Settings menu.

- Worries about the threat of biological terrorism have forced epidemiologists to make predictions about the spread of agents such as smallpox or even Ebola introduced to large metropolitan areas. Predictions prior to 2000 were based on the assumption that thousands would be infected to start an outbreak, while more recent models have been based on the threat posed by a single infected bioterrorist. The journal *Emerging Infectious Diseases* has frequent articles on the threat of bioterrorism. Use information you collect to generate models of the spread of a biological agent under different conditions and to evaluate the potential benefit of a vaccination program.

   Resources:

   Emerging Infectious Diseases (http://www.cdc.gov/ncidod/eid/index.htm)

- In the 1970s and 1980s big cities like San Francisco and New York attracted gay men from the rest of the country. This huge immigration of susceptibles may have fueled the AIDS epidemic. Assuming a 10% infection rate, and a high partner exchange rate, test the hypothesis that the influx of new susceptible men to these cities was necessary to create epidemic dimensions of this outbreak. (In *Epidemiology*, increase the birth rate to bring new susceptibles into the population.)

   Resources:

   Garrett, Laurie. *The Coming Plague*, 1994

- The 1918 Influenza epidemic had a major impact on the city of Philadelphia, which had a population of 1.5 million at that time. Influenza began with a few cases in Pennsylvania in September, 1918. The state case load increased to 75,000 cases before cases were first noted in Philadelphia. Influenza ravaged Philadelphia for three weeks and then virtually disappeared. Forty-seven thousand cases were reported and 12,191 people died. Use models of disease spread and of quarantine to explain explain the end of this outbreak, or track the progress of the epidemic in other areas.

   Resources:

   http://www.pbs.org/wgbh/amex/influenza/

   http://www.stanford.edu/group/virus/uda/

   "Maps and Microbes", Chapter 12 in *Microbes Count!*

- We are often told that sexually transmitted diseases (STDs) have reached "epidemic proportions." One city that is often cited as having a problem is Baltimore, Maryland, where changing social condition in the city led to the spread of syphilis from one set of neighborhoods to others. Malcolm Gladwell describes the Baltimore sexually transmitted outbreak in his popular book, *The Tipping Point*, as an example of how multiple conditions can contribute to extraordinary events. Review the epidemiology of STDs by looking at articles in *Morbidity and Mortality Weekly Reports (MMWR)* or *Emerging Infectious Diseases*, and develop models to explain spread and control strategies in a community of your choice.

   Resources:

   Centers for Disease Control (http://www.cdc.gov)

Software Used in this Activity

Epidemiology

Daniel Udovic and Will Goodwin (University of Oregon)

Platform Compatibility: Macintosh PPC

A web-based version of *Epidemiology* is also available. This is a beta version which is not yet complete, but it offers a good alternative where it is not feasible to use the Macintosh version. (http://darkwing.uoregon.edu/~bsl/epidemiology/index.html).

Additional Resources

Available on the *Microbes Count!* web site at http://bioquest.org/microbescount

Software

Epidemiology (Macintosh PPC version)
The "EbolaZaire" model file for *Epidemiology*

Text

A PDF copy of this activity, formatted for printing

Related *Microbes Count!* Activities

Chapter 9: Measles in Nakivale Refugee Camp: Epidemiology in Action

Chapter 9: A Plague on Both Houses: Modeling Viral Infection to Control a Pest Outbreak

Chapter 12: Maps and Microbes

Chapter 12: Vaccine: Experimenting with Strategies to Control Infectious Disease

Relevant Textbook Keywords

Endemic, Epidemic, Epidemiology, Immunity, Rhinovirus, Transmission (rate), Vaccine, Virulence

Related Web Sites (accessed 2/26/03)

A report on the 1918 influenza epidemic
http://www.stanford.edu/group/virus/uda/

Centers for Disease Control
http://www.cdc.gov

Centers for Disease Control, Divisions of HIV/AIDS Prevention
http://www.cdc.gov/hiv/pubs/facts.htm

Emerging Infectious Diseases, a jounal tracking and analyzing disease trends
http://www.cdc.gov/ncidod/eid/index.htm

Morbidity and Mortality Weekly Reports
http://www.cdc.gov/mmwr/

Program for Monitoring Emerging Diseases (ProMED)
http://www.promedmail.org

Public Broadcasting Service (PBS), The American Experience: *Influenza 1918*
http://www.pbs.org/wgbh/amex/influenza)

Understanding the Ebola Virus
http://www.cdc.gov/ncidod/dvrd/spb/mnpages/dispages/ebola.htm

References

Breman, J. G. , P. Piot, K. M. Johnson, M. K. White, M. Mbuyi, P. Sureau, D..L. Heymann, S. Van Nieuwenhove, J. B. McCormick, J. P. Ruppol, V. Kintoki, M. Isaacson, G. Van Der Groen, P. A. Webb, K. Ngvette (1977). The epidemiology of ebola haemorrhagic fever in Zaire, 1976. In *Proceedings of an International Colloquium on Ebola Virus Infection and Other Haemorrhagic Fevers*. Pattyn, S. R., editor. Prince Leopold Institute of Tropical Medicine, Belgium.
http://www.itg.be/ebola/ebola-24.htm

Gladwell, M. (2002). *The Tipping Point*, Back Bay Books: Boston.

Preston, R. (1995) *The Hot Zone*. Anchor Books: New York.

Udovic, D. and W. Goodwin (2001). Epidemiology. In *The BioQUEST Library Volume VI*. Jungck, J. R. and V. G. Vaughan, editors. Academic Press: San Diego, CA.

Bibliography

Garrett, L. (1994). *The Coming Plague: Newly Emerging Diseases in a World Out of Balance*. Farrar Straus & Giroux: New York.

Gordis, L. (2000). *Epidemiology*. WB Saunders Co.: Baltimore.

Meltzer, M. I., I. Damon, J. W. LeDuc,and J. D. Millar (2001). Modeling Potential Responses to Smallpox as a Bioterrorist Weapon. *Emerging Infectious Diseases*, Vol. 7, No. 6.
http://www.cdc.gov/ncidod/EID/vol7no6/meltzer.htm

Special Pathogens Branch, Centers for Disease Control, Ebola Hemorrhagic Fever
http://www.cdc.gov/ncidod/dvrd/spb/mnpages/dispages/ebola.htm

Figure and Table References

Figure 1.   Centers for Disease Control (http://www.cdc.gov)

Figure 2.   Screen shot from *Epidemiology*

Figure 3.   Screen shot from *Epidemiology*

Table 2.   Adapted from Figure 2 in Breman (1977)

Chapter 12

Activities for Video XII: Microbes and Human Disease

The continuing exploration of the complex relationships between microbes, environment, and disease has generated an array of information, methodologies, and tools. Continuously updated clinical and environmental data inform our health protocols and policies. New computational and visualization tools influence the way epidemiologists track outbreaks.

In this unit we can:
- weigh the personal risks of exposure to a potentially fatal virus,
- analyze maps of an outbreak and epidemiological data to characterize the disease at the population level, and
- design novel vaccine strategies and test their efficacy in population models.

It All Started Here

Starting with a case study based on the outbreak of a pulmonary disease in Southwest US, the interactions between human behavior, exposure, and confusing clinical symptoms are highlighted. As we search for the cause, these complications reflect real life difficulty in making diagnoses. The real questions extend beyond what disease has been contracted to cascading effects on the society we live in.

Maps and Microbes

Mapping the spread of a disease is no trivial task. Using maps from the Centers for Disease Control (CDC) and the United Nations Progromme on AIDS (UNAIDS), we can look for patterns of spread and test hypotheses about disease causation at the population level. We can contrast the spread of current outbreaks like West Nile virus to that of the epidemic spread of influenza in 1918. Geographical Information Systems (GIS) prove to be a powerful tool for analyzing local disease outbreaks as well.

Vaccine: Experimenting with Strategies to Control Infectious Disease

Models are often used as decision-making tools. We can use the program Epidemiology to model the effects of vaccines on the HIV epidemic in a severely affected community. We can contrast the impact of different vaccines and alternative strategies

Souvenirs: Investigating a Disease Outbreak

Janet Yagoda Shagam, Ethel D. Stanley, and Janet M. Decker

Video XII: Microbes and Human Diseases

Souvenirs creates a problem space involving disease, multiple characters, and the realistic situations they face. As you attempt to untangle the skein of information in this scenario, you will have the opportunity to gain a practical level of understanding for the science needed to solve the problem. In addition to the microbiological investigation, there are threads involving social, economic and political impacts of disease on a community.

Getting Started

Choose someone in your group to read the first part of *Souvenirs* out loud. What do you think the case is about? Work with your group to answer the first set of questions. Feel free to use the resources available in the lab including your textbook and access to the internet. (The last three parts of the case are on the web site and will be assigned independently.)

Part 1. It All Started Here

It had been a wonderful week and a terrific break from the hum-drum life-style of the Chicago suburbs. The three women, all best friends from their college days, had planned for over a year to go on this archaeological dig sponsored as a fund raiser by their alma mater. Hilary, an elementary school teacher who suffered from asthma, was a bit concerned about working at such a high altitude. Ruth, a pharmaceutical chemist, was relieved to get away from her non-responsive colonies of laboratory mice. For Susan, this was just a bit of much needed R and R from her hectic life as a commodities broker. Although the dig had dried to dust, the participants stopped often to admire the yellow, pink and blue flowers covering the nearby hillsides.

The pot shards they had collected, carefully numbered and boxed, were now ready to be sent on to the museum. The friends placed the boxes, along with the plant material collected by the ethnobotanist, in the old storage shed at the site. Then they took the jeep for the one hour drive down the mountain to Santa Fe, where they spent their last afternoon in the Southwest shopping for jewelry and rugs. Because they had skipped lunch, they bought some fruit from a corner grocer. He explained in halting English that his family had just moved to Santa Fe from Argentina to start a new life. The fruit looked and tasted wonderful, though the old man coughed terribly as he waited on them. "I hope that's not catching!" Ruth thought to herself. Then she noticed a half-empty pack of cigarettes near the cash register and forgot the incident.

Figure 1: The kiva, a round chamber built underground with entry by a ladder, is a unique archaeological feature found in the southwestern United States.

The next morning was hectic, with last minute packing and loading the truck. As a favor to the site-manager, Hilary and Ruth also cleaned the storage shed. It was apparent that the shed was home to many rodents and had not been cleaned for many months. Hilary and Ruth did the best they could in the time they had. Both hoped someone else would complete the cleaning before the next group arrived to work at the site.

The flight back to O'Hare was smooth, but crowded. Susan and Ruth sat together, but Hilary sat in the back so she could stretch out a bit. Although the leg room was nice, it was annoying to sit right under the cabin air vent and in such close proximity to the bathrooms. Hilary cleared her throat.

Now ask someone else to read the second part.

Part 2. Home Sick

Sunday in Chicago for the three women was uneventful. They spent their time collecting the mail from the neighbor kid and catching up on laundry and bills before going back to work the next morning. Hilary had bought a beautiful hand-woven Navajo rug, which she shook out and displayed to family and friends.

Hilary began to feel ill late Sunday afternoon, with a fever, chills, and a dry cough. Since a cough was often a prelude to a long bout of asthma, Hilary used the nebulizer to insure a good dose of anti-inflammatory and bronchial dilating medications. She was also very thirsty. "It must be the result of my dry week in the Southwest and the canned airplane air," she thought to herself.

The next morning, feeling somewhat better, Hilary went off to work. It would be good to see "her" third grade kids after their spring break. However, by first recess Hilary was gasping for air. She was taken to the hospital and a substitute teacher took her place while the kids were out on the playground. Hilary was pronounced dead at 7:28 Monday evening. Cause of death was listed as respiratory failure due to asthma.

Susan, whose morning always started with obituaries and coffee, screamed as she read Tuesday's paper. She immediately called Ruth. The phone rang 6 times. "Hello?" croaked Ruth.

"Ruth, have you seen the paper? Hilary died last night!" Susan sounded breathless, as if she had been running.

"Hilary...but we just saw her Saturday and she was feeling fine!" answered Ruth.

"You don't sound so great yourself... Are you O.K.?" asked Susan.

"Yeah, I just caught a cold on that flight from Albuquerque. When's the funeral? I can't believe this is happening."

- Identify phrases having to do with any respiratory disorders found in the characters in the case?

- What are some possible non-microbial causes of the respiratory disorders?

- What kinds of microorganisms can cause respiratory disease?

- Identify sources of microbes that could be implicated in a respiratory problem for these characters.

Now that you have gathered some preliminary information, continue with "Part 3: What Hit Us?" on the *Microbes Count!* web site.

Optional Individual Activity

In the situation below, you asked to make a decision that requires you to consider biological information that you may not be familiar with. Read the scenario and gather facts. You will then be asked to make and defend your personal decision.

The Job Dilemma

You're an Anthropology major with a minor in Museum Studies. You've hoped to work in the US with Native American and early Spanish artifacts in the Southwest, but the field is tight for college grads in this area- most of the other people competing for jobs have at least a master's degree.

You've been offered the job of your dreams at a small museum and archeological site in southwest Colorado. The museum is old and musty, but you will be working on cleaning it up and rearranging the displays. Some on site digging for artifacts is planned as well. The museum director has warned you that the museum is in the middle of the area where deer mice are frequently found and that for much of the reorganization work in the museum, you'll be required to wear a respirator and oxygen tanks.

Should you accept? You need to make a decision.

- Make a list of at least 10 facts that will guide your decision. For each fact, be sure to specify the source, e.g. a book like "The Coming Plague" (Garrett 1994), a website like CDC (http://www.cdc.gov/ncidod/dvrd/revb/respiratory/eadfeat.htm), etc.

- Will you go? Explain.

Additional Resources

Available on the *Microbes Count!* web site at http://bioquest.org/microbescount

Text

A PDF copy of this activity, formatted for printing

Souvenirs Part 3: What Hit Us?

Souvenirs Part 4: Of Mice and Men

Souvenirs Part 5: Loose Ends and Law Suits

Medical data for Hilary, Ruth, Grocer, and Site Manager

Related *Microbes Count!* Activities

Chapter 5:   The Farmer and the Gene: A Case Approach to Bt Corn

Chapter 10: Measles in the Nakivale Refugee Camp: Epidemiology in Action

Chapter 11: Identifying Immune Responses

Unseen Life on Earth Telecourse

Coordinates with Video XII: Microbes and Human Disease

Relevant Textbook Keywords

Asthma, Infection, Respiratory

Related Web Sites (accessed on 4/20/03)

Microbes Count! Website
   http://bioquest.org/microbescount

Respiratory System: Diseases.
   http://home.cc.umanitoba.ca/~eworobe/122lecture15.pdf
   http://oac.med.jhmi.edu/res_phys/DiseaseStates/index.html
   http://www.cdc.gov/ncidod/dvrd/revb/respiratory/eadfeat.htm

Respiration System
   http://science.nhmccd.edu/biol/ap2.html#respir

Unseen Life on Earth: A Telecourse
   http://www.microbeworld.org/htm/mam/is_telecourse.htm

References

Fass, M. (2002) *The Job Dilemma*. (Private Communication)

Garrett, L. (1994). *The Coming Plague: Newly Emerging Diseases in a World Out of Balance*. Penguin Books:New York.

Figure and Table References

Figure 1.   Courtesy Janet Yagoda Shagam

Maps and Microbes

Marion Field Fass, Amy Sapp, and Janet Vigna

Video XII: Microbes and Human Disease

The spread of microbial pathogens is determined by the availability of niches in which they can survive and thrive. The spread of HIV/AIDS, the West Nile virus, and the flu epidemic of 1918 have demonstrated that for microbial pathogens, access to niches is created by geography, ecology and host behavior. These factors, combined with host susceptibility, determine the epidemiology of a disease. New technologies, such as Geographical Imaging Systems (GIS), can expand our understanding of infectious diseases.

In using GIS to track pathogens and to explore the factors that influence their spread, the researcher goes through three steps:

Figure 1. Fighting flu in 1918.

Step 1. Surveillance

- What is the current distribution of this disease? That is, where do the infected hosts reside?

- What is the current distribution of disease vectors, if relevant?

- What other organisms, besides humans, might be involved in the disease process?

Step 2. Historical perspective

- What was the geographical distribution of this disease among hosts (and vectors, if relevant) in the past?

- How have these distributions changed over time? Try to look back at least 100 years.

Step 3. Hypothesizing spatial correlates of pathogen distribution

- What research has already been conducted? What variables were included?

- What variables are absent from existing studies?

Scientists can hypothesize about many different types of variables—from factors that affect health at the individual level, such as genetics, to those that affect health at the global level, such as poverty.

In thinking about what variables to include in a new study, scientists should first find out what variables already have been studied, and what variables have not been studied or have not been studied adequately. To elucidate these variables, scientists employ *theories* of disease and its distribution. These theories, which are composed of structured sets of ideas about the relationships between concepts, provide a framework for hypothesizing about factors that influence whether individuals or populations will become infected with a disease.

The Biomedical Model

For example, the biomedical model is a theory that a researcher might use to hypothesize about individual-level "risk factors," which can be individual behaviors, exposures to tangible hazards, or biological conditions. The biomedical model seeks to understand how these "risk factors" contribute to disease mechanisms, usually mechanisms at the cellular level. The biomedical model is more concerned with disease causation than with disease distribution.

A scientist using the biomedical model to formulate hypotheses about determinants of pathogen infection, then, would think about how individual behaviors, exposures and conditions might place people at risk, and the biological mechanisms through which these factors render people susceptible to disease. For example, individuals with impaired immune systems might be more susceptible to pathogen infection because their immune systems cannot fight off invasive pathogens at the same rate as individuals with normal immune systems.

The Ecosocial Theory

A researcher might use ecosocial theory to formulate hypotheses about why the distribution of disease is unequal within and among different populations. It is a theory concerned with disease distribution as opposed to disease causation. Ecosocial theory considers not only individual-level risk factors, but also factors that affect health on a variety of levels over time. The central question this theory asks is, "Who and what is responsible for population patterns of health, disease, and well-being, as manifested in present, past, and changing social inequalities in health?" Ecosocial theory calls for examination of how socioeconomic position, gender, race/ethnicity, sexuality, age, immigrant status and other social categories act on individual biology to influence disease susceptibility or vulnerability (Kreiger, 2001).

For example, in the case of tuberculosis (TB) in Russia, a researcher using ecosocial theory might examine the distribution of infection in different socioeconomic groups to find that infection rates are highest among the poorest populations. The researcher would examine factors that might explain TB's unequal socioeconomic distribution, considering the role of material factors in shaping exposure, infection, and the course of disease development. The researcher might conclude that poor people have a greater risk of being exposed to TB because they live and/or work in overcrowded, poorly ventilated areas; they are more likely to develop TB infection, once exposed, because they cannot afford adequate nourishment and consequently have weakened immune systems; and they might be more adversely affected by TB, once infected, because they lack access to treatment such as drug therapy.

The theory behind hypothesis formulation determines what a scientific study will reveal, and what it will hide. Being explicit about theory should help researchers, and those that rely on research, to understand that science cannot be completely objective.

Using GIS

Geographical Information Systems (GIS) have become very useful tools for microbiologists and epidemiologists. In this activity we will track the spread of

HIV/AIDS, study the spread of West Nile Virus, and explore the path of infection with influenza in 1918. We can formulate hypotheses and test them using available data. For further activities, students can explore the distribution of many other diseases through maps easily available though the Centers for Disease Control and other agencies. (For best results, search through Google.com, Image search. Search using the disease and the word "map" or "maps", for example, "malaria maps".)

I. Exploring the spread of HIV/AIDS

Maps allow us to ask questions about the factors that influence disease spread. The changes in distribution of HIV in the world have demonstrated that the increase in cases of disease is not a linear and constant process. The maps in Figures 2-6, which show the distribution of AIDS in the world and the spread of HIV/AIDS through Sub Saharan Africa over the last two decades, raise questions about vulnerability to HIV/AIDS.

(Larger, color versions of all of the maps in this activity are available on the *Microbes Count!* site. In addition, more current versions of some of the maps may be available from the websites listed in the Figure and Table References section at the end of this activity.)

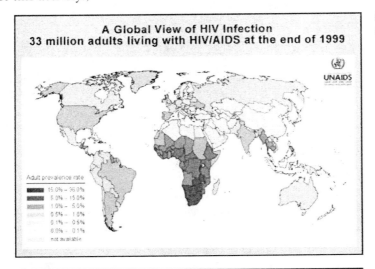

Figure 2. Percent of adults with HIV/AIDS at the end of 1999.

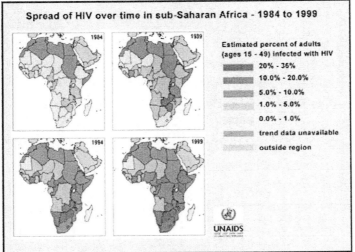

Figure 3. Spread of HIV in sub-Saharan Africa between 1984 and 1999. (Please note: the countries of northern Africa are out of the region measured. The prevalence in these countries is lower than it appears in this black and white map. Please check the color map available on the web site.)

Marion Field Fass, Amy Sapp, and Janet Vigna   The BioQUEST Curriculum Consortium

1. Look at the map of the distribution of cases of HIV/AIDS in the world in 1999 in Figure 2.
   - Describe the differences in prevalence that you observe.
   - What hypotheses can you make about differences observed?

2. HIV was first identified in both central Africa and in the United States in the early 1980s.
   - How do the prevalence levels in these areas compare now?
   - What factors can you suggest that might differentiate the establishment of disease in these different populations?

3. Look at the maps of the spread of HIV in African countries from 1984 through 1999 in Figure 3.
   - Is HIV evenly distributed in the region?
   - Is the rate of change in the percent of infection the same in every country?
   - What factors can you think of that might be responsible for the differences in the rate of change?
   - What data do you need to test your hypotheses?

4. The graphs of HIV prevalence in Kenya, Uganda and South Africa in Figures 4, 5, and 6 complement the data presented in the maps above.
   - What questions can you generate from these graphs about changes in prevalence over time in these three countries from the same region?
   - What research would you need to do to answer your questions about differences in prevalence in these neighboring countries?

Figure 4. Estimated prevalence of HIV in adults in Kenya between 1990 and 1999.

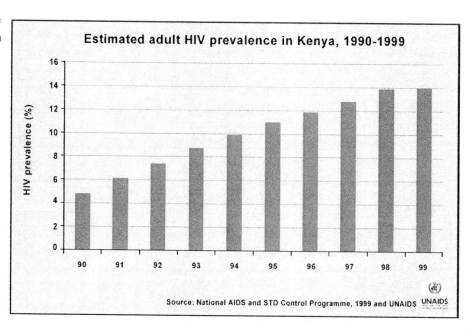

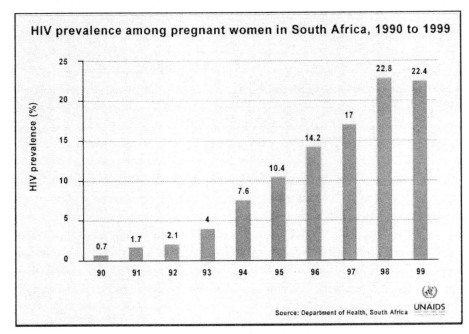

Figure 5. HIV prevalence among pregnant women in South Africa, 1990 to 1999.

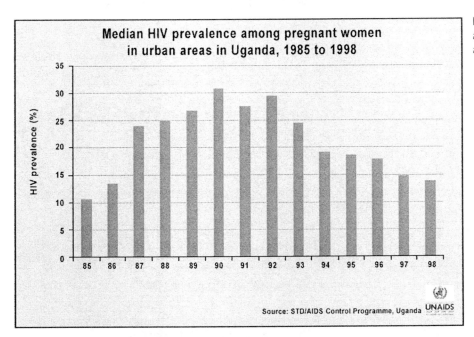

Figure 6. Median HIV prevalence among pregnant women in urban areas in Uganda, 1995 to 1998.

II. Planning for West Nile Virus

In 1999 the first cases of West Nile Virus infection were noted in the New York City area in both birds and humans. Prior to this time West Nile virus had only been recognized in Africa, Europe and the Middle East. West Nile virus is carried by Culex mosquitoes and most frequently infects birds and horses. Humans are only incidental hosts.

The map in Figure 7 illustrates the spread of West Nile Virus by state from 1999 through 2002, and notes the states in which human infections have been confirmed.

Figure 7. Spread of West Nile Virus by state, 1999-2002. West Nile Virus Activity in the U.S. in birds, horses mosquitoes, animals, or humans.

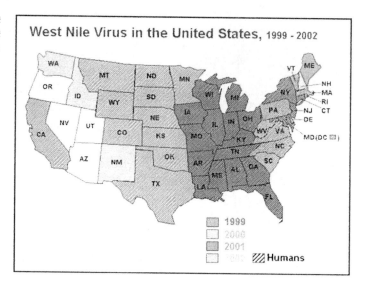

1. How can you describe the pattern of movement of this virus in the US?
   - What are the advantages of using states as the level of analysis?
   - What are the disadvantages?

2. West Nile virus has obviously become established in mosquito populations across the United States. Think about the life cycle of the virus and the disease to answer these questions.
   - Do you think that cases in humans will increase? Why?
   - What factors might limit the number of human cases?
   - What factors have led to its successful spread throughout the US?
   - What is your hypothesis based upon?

Figure 8. Human cases of West Nile Virus infection as of January 17, 2003.

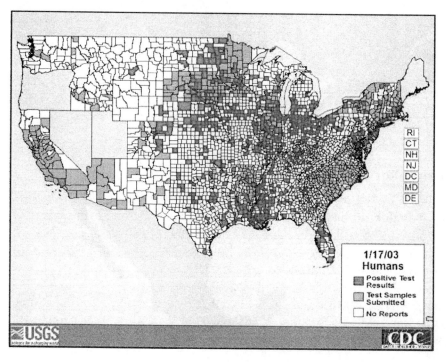

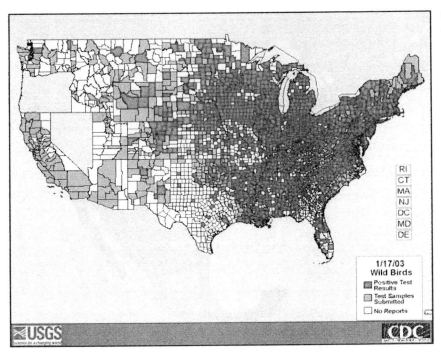

Figure 9. Cases of West Nile virus infection in wild birds as of January 17, 2003.

3. The map in Figure 8 looks at human cases of infection with West Nile virus whereas the map in Figure 9 documents cases found in wild birds. Both distributions are shown by county.

- Contrast the distribution of diseases in these two organisms.

- Discuss the value of studying the infections in birds as well as people as microbiologists attempt to plan for future control efforts.

III. Understanding the impact of the 1918 Influenza epidemic

The Influenza pandemic of 1918 killed more people in the United States in one year than were killed in combat during all of the wars of the 20th century combined.

While a graph such as Figure 10 can detail the changes in death rates, maps such as those in the maps in Figures 11a - 11e provide another layer of information about the nature of the epidemic.

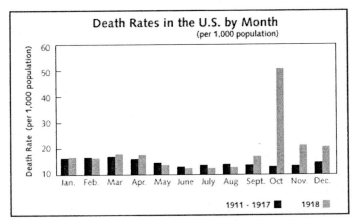

Figure 10. Death rates in the United States in 1911-1917 and in 1918.

Figure 11a-e. Spread of the influenza epidemic, September 14 through October 5, 1918.

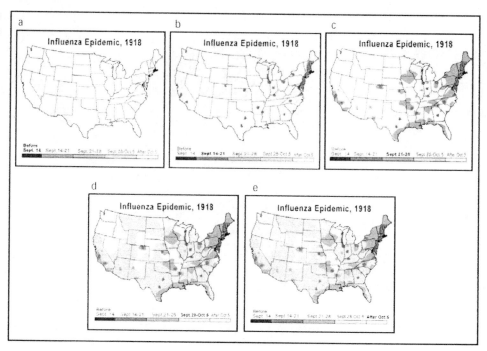

1. What hypotheses can you make about the factors that influenced the spread of this epidemic?

2. How did it move?

3. Why did it end?

IV. Creating your own maps

You can create maps using GIS software with your own data. A sample file with data on 1918 Influenza epidemic is included on the *Microbes Count!* site. This file contains information about the arrival of influenza in various cities in the United States and the death rates in these cities before and during the influenza epidemic. If you have access to a GIS lab, these data can be plotted in map format.

You can also search on the web for additional data on the flu epidemic, HIV/ AIDS, and West Nile Virus, as well as many other topics. Some good sources are listed in the Additional Resources: Web Sites section at the end of this activity.

Software Used in this Activity

ArcExplorer

ArcExplorer is a free GIS viewer available from ESRI. *ArcExplorer* can be used to view and manipulate maps created using the commercially available *ArcView application*. It can be downloaded from the ESRI web site at: http://www.esri.com/software/arcexplorer/overview.html

Platform Compatibility: *ArcExplorer* is available in three versions (information current as of February 27, 2003):

ArcExplorer2 (Windows Edition): Windows 98/2000/NT/XP
ArcExplorer4 (Java Edition): Windows, UNIX, or Linux
ArcExplorer Web: Standard web browser

Additional Resources

Available on the *Microbes Count!* web site at http://bioquest.org/microbescount

Text

A copy of this activity, formatted for printing

Color versions of all of the maps

Data on the 1918 Influenza epidemic

Related *Microbes Count!* Activities

Chapter 12: Vaccine: Experimenting with Strategies to Control Infectious Disease

Unseen Life on Earth Telecourse

Coordinates with Video XII: Microbes and Human Disease

Relevant Textbook Keywords

Epidemiology, Incidence, Prevalence

Related Web Sites (accessed on 3/23/03)

Centers for Disease Control
http://www.cdc.gov

Joint United Nations Programme on HIV/AIDS
http://www.unaids.org

Microbes Count! Website
http://bioquest.org/microbescount

PBS, The American Experience: *Influenza 1918*
http://www.pbs.org/wgbh/amex/influenza

Unseen Life on Earth: A Telecourse
http://www.microbeworld.org/htm/mam/is_telecourse.htm

World Health Organization
http://www.who.int/en/

References

Crosby, A. (1990). *America's Forgotten Pandemic: The Influenza of 1918.* Cambridge University Press: New York.

Krieger, N. (2001). Theories for social epidemiology in the 21st century: an ecosocial perspective. *International Journal of Epidemiology* 30:668-677, http://www.ccs.ufsc.br/geosc/socialtheories.pdf

Figure and Table References

Figure 1.   http://www.cdc.gov/ncidod/diseases/flu/viruses.htm

Figure 2.   Courtesy UNAIDS. *Report on the global HIV/AIDS epidemic - June 2000.* Slides in PowerPoint: Slide 14 http://www.unaids.org/epidemic_update/report/epi-core.ppt

Figure 3.   Courtesy UNAIDS. *Report on the global HIV/AIDS epidemic - June 2000.* Slides in PowerPoint: Slide 16 http://www.unaids.org/epidemic_update/report/epi-core.ppt

Figure 4.   Courtesy UNAIDS. *Report on the global HIV/AIDS epidemic - June 2000.* Slides in PowerPoint: Slide 4 http://www.unaids.org/epidemic_update/report/epi-ext.ppt

Figure 5.   Courtesy UNAIDS. *Report on the global HIV/AIDS epidemic - June 2000.* Slides in PowerPoint: Slide 3 http://www.unaids.org/epidemic_update/report/epi-ext.ppt

Figure 6.   Courtesy UNAIDS. *Report on the global HIV/AIDS epidemic - June 2000.* Slides in PowerPoint: Slide 1 http://www.unaids.org/epidemic_update/report/epi-ext.ppt

Figure 7.   http://www.cdc.gov/ncidod/dvbid/westnile/surv&control.htm

Figure 8.   http://cindi.usgs.gov/hazard/event/west_nile/west_nile.html

Figure 9.   http://cindi.usgs.gov/hazard/event/west_nile/west_nile.html

Figure 10.   Adapted from Crosby, A. (1990). *America's Forgotten Pandemic: The Influenza of 1918.* Cambridge University Press: New York.

Figure 11.   Adapted from Crosby, A. (1990). *America's Forgotten Pandemic: The Influenza of 1918.* Cambridge University Press: New York.

Vaccine: Experimenting with Strategies to Control Infectious Disease

Marion Field Fass

Video XII: Microbes and Human Disease

Vaccines for HIV are being tested, but no one expects that HIV will disappear immediately. Vaccines against HIV may not offer complete protection, and will not be immediately available in sufficient numbers to vaccinate everyone who is potentially at risk of contracting HIV/AIDS. Consumer acceptance of the vaccine may also vary because side effects without complete protection may foster rejection. Given these realities, it will be critically important to forecast how an immunization program will affect disease spread and what will be the best way to distribute limited supplies of the vaccine.

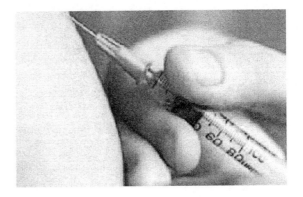

At the population level, morbidity and mortality from HIV/ AIDS are influenced by:

Figure 1. Injecting a vaccine.

- how many people become infected with the virus,

- how quickly people who are infected progress to the end stage of infection, and

- how likely it is for a sick individual to transmit the virus to a healthy partner.

Your challenge is to figure out how to distribute a new vaccine to a population at risk. First let's consider strategies for using the perfect vaccine.

The Perfect Vaccine

Let's think first about what characteristics a perfect vaccine would have:

- What are the attributes of the perfect vaccine?

- How do the vaccines that are commonly used for preventing infectious diseases measure up to these standards?

- Explore the characteristics of the vaccines for influenza, polio, smallpox, measles.

- What are their advantages? What are the risks in using them?

- What would be the "perfect" vaccine for HIV/AIDS? Why?

Using a Great Vaccine

Suppose that a vaccine has been discovered that is 100% effective at preventing infection with HIV after sexual exposure. Consider the following scenario:

As a health officer in a community where HIV is primarily spread by heterosexual sex, you have received a supply of vaccine. Your allocation for the first year is adequate to vaccinate only 10% of the population over 15 years of age. Already you've received phone calls and letters from the Mayor and members of the City Council requesting the vaccine for themselves and their families. Other calls from "important people" are waiting.

You can assume that in your community:

- 50% of the adult men are at high risk because of multiple partners,
- 50% are at low risk,
- younger women (<20) may have multiple partners and account for 10% of the population,
- one percent of the women are commercial sex workers, and
- other women are most at risk of being infected by their husbands.

- What strategy would you use to determine what would be the most effective way to distribute this vaccine?

Define your community in terms of which groups are most at risk of infection and which groups stand to benefit from vaccination. Contrast the spread of HIV in the community if no vaccine is delivered and then if the vaccine is delivered to the highest risk groups.

You can use the program *Epidemiology* to predict the response of your population to vaccination. You will need to perform separate runs for high and low risk groups in the community. (An installer for Epidemiology is included in the Vaccine folder on the *Microbes Count!* web site. Information on using *Epidemiology* is available in the Users Manual on the *Microbes Count!* web site. If you have not used Epidemiology before you may also want to try the Epidemiology activity "Epidemiology: Understanding Disease Spread" in Chapter 11.)

For real data about HIV incidence and prevalence, consult the Centers for Disease Control (CDC) website for United States data, and the Joint United Nations Program on HIV/AIDS (UNAIDS) website for data from other countries (see the Additional Resources section at the end of this activity.) Your vaccine distribution plan will benefit from the use of concrete examples.

The Other Vaccine

The perfect vaccine may be only a dream. Some experts have suggested that the spread of HIV could be significantly reduced by vaccines that are less than perfectly effective. A vaccine may be only 25% effective at preventing someone who has been vaccinated from acquiring HIV if exposed, but it may also affect the progression of the infection to disease in already infected people, and decrease the viral load in their body fluids. This would reduce the probability that they would pass the virus to a partner during unprotected sexual intercourse. This vaccine might have the same effect on the incidence of new cases of HIV as the adoption of Highly Active Anti-Retroviral Therapy had in the United States in the mid-1990s.

One can imagine a vaccine that is 75% effective in reducing the progression of the infection with HIV, so that while people may remain infected, they do not become ill as rapidly. In addition, the vaccine might also be 75% effective in reducing the probability of transmission of the virus from one individual to another.

Using a Less Than Perfect Vaccine

As a health officer in the community described above you again receive only enough vaccine for 10% of the population this year. You know that in this country 15% of adults are HIV positive.

What are the parameters you should consider in deciding how to use your vaccine? First you need to predict how the epidemic will progress without the vaccine. As before, assume that 50 % of the adult men are at high risk because of multiple partners, 50% are at low risk, younger women (<20) may have multiple partners and account for 10% of the population, one percent of the women are commercial sex workers, and other women are most at risk of being infected by their husbands.

Using *Epidemiology*, consider each time unit a month, and run your simulations for 120 months.

- How will you adapt your model to account for different behaviors in different risk groups? As before, make different models for different groups in the population.

- According to your model, how many people will be infected without treatment at the end of 120 months?

- When the vaccine arrives you must decide what is the best way to distribute it? Model the effects on the population outcome in 120 months of different distribution schemes. Make sure to adapt your model for different sub-populations.

- What are your recommendations for the most effective distribution scheme?

This scenario is based on a presentation entitled "Optimal HIV vaccination policies in developing countries: The importance of indirect vaccine effects in limiting HIV transmission" by Professors Daniel Barth-Jones and Ira Longini of Wayne State University Medical School at the American Public Health Association Annual Meeting, Atlanta Georgia, October 24, 2001. The work of Drs. Barth-Jones and Longini was supported by the CDC and UNAIDS.

Software Used in this Activity

Epidemiology

Daniel Udovic and Will Goodwin (University of Oregon)

Platform Compatibility: Macintosh

Additional Resources

Available on the *Microbes Count!* web site at http://bioquest.org/microbescount

Software

Epidemiology

Text

A PDF copy of this activity, formatted for printing

Related *Microbes Count!* Activities

Chapter 9:   Measles in Nakivale Refugee Camp: Epidemiology in Action

Chapter 9:   A Plague on Both Houses: Modeling Viral Infection to Control a Pest Outbreak

Chatper 11: Epidemiology: Understanding Disease Spread

Chapter 12: Maps and Microbes

Unseen Life on Earth Telecourse

Coordinates with Video XII: Microbes and Human Disease

Relevant Textbook Keywords

HIV, Immunity, Prevention, Public health, Vaccination, Vaccines

Related Web Sites (accessed 2/26/03)

Centers for Disease Control
http://www.cdc.gov

Microbes Count! Website
http://bioquest.org/microbescount

The Joint United Nations Programme on HIV/AIDS (UNAIDS)
www.unaids.org

Unseen Life on Earth: A Telecourse
http://www.microbeworld.org/htm/mam/is_telecourse.htm

References

Barth-Jones, D. and I. Longini (2001). Optimal HIV vaccination policies in developing countries: The importance of indirect vaccine effects in limiting HIV transmission. Presentation given at the American Public Health Association Annual Meeting, Atlanta, Georgia, October 24, 2001.

Udovic, D. and W. Goodwin (2001). Epidemiology. In *The BioQUEST Library Volume VI*. Jungck, J. R. and V. G. Vaughan, editors. Academic Press: San Diego, CA.

Figure and Table References

Figure 1.   www.rush.edu/happening/hiv-vaccine.html

Microbes Count!

Celebrations and Challenges

Celebrations

Microbes Count! activities emphasize problem posing, problem solving and peer persuasion in microbiology. The members of the BioQUEST Curriculum Consortium who have contributed to this believe that these processes are central to any scientific investigation. We hope that the investigations you have encountered have engaged and challenged you. We encourage you to use and adapt the tools, resources, and methodologies introduced in *Microbes Count!* to investigate new problems that interest you. Your experiences with video microscopy, image analysis, modeling and statistical packages, and geographic systems are likely to be useful in biological problems you encounter outside of *Microbes Count!*

The American Society for Microbiology's *Microbial Literacy Collaborative* initiative in microbiology embraces contemporary research in microbial physiology, ecology, and evolution and genetics. By developing complementary investigations to accompany *Microbes Count!* extends new opportunities to explore the richness of contemporary microbiology as a discipline. Not only can you study the application of microbiological research to problems associated with human health, our environment, and agriculture, but also you can examine the relationship between contemporary human society and the microbiota on earth.

As you have read and used the activities in *Microbes Count!*, you have had the opportunity to explore some of the most famous problems in microbiology. Indeed, their "solutions" have lead to awards such as the Nobel Prize in Physiology and Medicine. What took a team of investigators years to accomplish in the past, researchers (including students) now have the technology to duplicate and possibly extend in a single work session in the lab or at the computer. Such is the nation of scientific research that the "solution" leads to new questions and the development of new technologies.

Many of the investigations included in *Microbes Count!* are open-ended, so you could generate questions that could be explored for years. If you feel that you have a generated a fundamentally new insight as a result of one of these investigations, we urge you to think about submitting a research article to the *American Journal of Undergraduate Research* (a hard copy journal with faculty reviewers), the *Journal of Young Investigators* (a peer-reviewed on-line journal published and edited completely by undergraduates and recent graduates), or to one of many microbiology research journals such as the *Journal of Bacteriology*.

Challenges

Obviously *Microbes Count!* is a finite, temporal resource. Only a few of the areas of investigation currently being explored by microbiologists or the tools that they are using are included. We invite you and other users of *Microbes Count!* to add to this collection in future editions or on our web site: *bioquest.org/microbescount*.

In fact, numerous new investigations and tools that do not appear in the text are accessible online. For example, we invite you to explore *bioquest.org/esteem* which features a simulation developed by Anton Weisstein. It focuses on Luria and Delbrück's Nobel Prize-winning experiment on fluctuation analysis and allows you to explore the determination of mutation rates and the effect of evolutionary selection on bacterial populations. Several microbiology cases are featured at bioquest.org/icbl/cases.php on the new Investigative Case Based Learning site. You will also find on the *Microbes Count!* site an associated laboratory experiment by Douglas Green and Donna Balzone reprinted from the American Biology Teacher. Even more supplementary materials on the web site will allow you to extend the initial investigation.

We encourage you to investigate additional possibilities for quantitative problem solving, open-ended investigation, and collaborative research. Our short list includes: Quorom sensing and biofilm formation and function; the impact of being cell size (that is, operating at low Reynold's numbers in a viscous and often turbulent three dimensional environment); dynamic behavior of moving cells via flagella and/or cilia or via amoeboid motion (analysis of motility and chemotaxis); assembly of viral capsids (symmetry groups, T-numbers); X-ray crystallography of 3 D structure of molecules, assemblies such as antigen-antibody complexes, ribosomes, and viral capsids; signaling; data mining of DNA microarrays to explore differentiation, pattern formation, and morphogenesis; phylogenetic probing applied to problems such as bioremediation and identifying unculturable infectious microorganisms by metagenomics; or artificial life models of immunity which are important in areas like bioterrorism.

Finally, we hope that *Microbes Count!* will prepare you to take on the challenges of our era with the emergence of new infectious diseases such as SARS or avian flu and to deal with global problems of safely feeding the world, cleaning up our environment, addressing endemic and persistent infectious diseases that affect millions such as malaria, tuberculosis, dysentery and cholera, developing sustainable use of biofuels, and living safely with one another.

John R. Jungck
Beloit College